Praise for *The Battle Beyond*

"*The Battle Beyond: Fighting and Winning the Coming War in Space*, provides a superb foundation for the theory, doctrine, strategies, and tactics needed to deter—and, if necessary, to fight and win space wars. Starting with the timeless observations of Sun Tzu, and building on US current military doctrine, this book provides powerful 'big ideas' for operations in space. It includes thoughtful discussions on space principles of war, orbital choke points, linkages between the terrestrial and space conflict escalation ladder, space command and control, and strategic considerations for potential space warfare. Surprisingly, *The Battle Beyond* is also a great read that will be very valuable for both casual observers and professional space warfighters alike."

General David Petraeus, US Army (Ret.),

Former Commander of the Surge in Iraq, US Central Command, and NATO and US Forces in Afghanistan, and former Director of the CIA

"*The Battle Beyond: Fighting and Winning the Coming War in Space* by Paul Szymanski and Jerry Drew provides novel insight into the unique complexities of the space domain and space warfare concepts. It illustrates the inherent complexity of the space domain and supports the need for deliberate resource allocation, training, and operational concepts to ensure our future joint and combined arms operations are supported with robust space capability. Provides good perspective on the importance of preparing for war in space."

Timothy J. Harp,

Former Deputy Assistant Secretary of Defense for C4ISR, IT, Intelligence and Space; US Mission to NATO Director, Brussels, Belgium; and a Board member on five major NATO standing boards and committees.

"*The Battle Beyond* is an impressive work that can inform our space warfighters, strategists, and planners at a very critical time in our participation in on-going complex, multi-domain operations in multiple theaters. The evolving great power struggles we see today demand thoughtful approaches to our National Defense and Security Strategies...Paul and Jerry provide that."

Gene Renuart, General, United States Air Force, (ret.),
Former: Commander of NORAD – NORTHCOM; Senior Military
Assistant to the Secretary of Defense; Director of Strategic Plans
and Policy, The Joint Staff; Vice Commander of Pacific Air Forces.

"A terrific and necessary read for understanding all elements of deterring, fighting and winning in the Space Domain. Professional war fighters, put this on your shelf of must-reads. Highly recommended!"

Admiral Bill Gortney, United States Navy, (ret.),
Former: Commander of NORAD – NORTHCOM; Commander
of US Fleet Forces Command; Director of The Joint Chiefs
of Staff; Commander of US Fifth Fleet; and Commander of
Carrier Strike Group Ten, Harry S. Truman Strike Group.

"Our warfighter needs every advantage in future wars when engaging our enemy and the US needs every tool available to be win. *The Battle Beyond: Fighting and Winning the Coming War in Space* provides a perspective of war fighting in the outer space domain. It should be required reading by anyone interested into future conflicts and military strategy to secure an advantage over the enemy."

Rear Admiral Gregory J. Slavonic (ret.),
Former: Acting Under Secretary of the Navy; 18th Assistant
Secretary of the Navy; Chief of Staff of US Senator James
Lankford and US Congressman James Lankford; and, Director
of Public Affairs, Multi–National Force – Iraq (MNF–I).

"Paul Szymanski and Jerry Drew in *The Battle Beyond: Fighting and Winning the Coming War in Space* is an interesting read. It defined the audience and articulated relevance of the space environment for warfare to the military practitioner and to members of war colleges The authors' expertise shows, and the work put into the subject is enlightening and extensive. Major takeaways given in this book of significance:

If governments are not understanding of the space environment or not decisive in the employment of space forces, the effectiveness of military operations—and by extension, the attainment of policy objectives—may suffer.

One may choose to approach space warfare through a variety of lenses— policy, technical, institutional, ethical, and warfighting, to name a few. While these lenses are certainly interrelated, the primary audience for this discussion is the military practitioner.

The planning and execution of operations— regardless of the domain— requires visualization, especially for the three-dimensional complexities of outer space, as significantly illustrated in this book.

Must combine tactical actions to develop operational concepts or more holistic courses of action. Command and control in, from, and through space is a significant hurdle in this endeavor because—although there are analogues in other-domain operations—the space domain poses unique challenges that, like operations in the cyber domain, are outside the common experience of most military personnel.

Critical questions of military action begin and end with strategy and not necessarily tactical approaches.

Accomplishment of military operations require intelligence collection, communications, cyber activity, aerial infiltration, operations in urban terrain, exfiltration, and intelligence exploitation. Space provides support to all of these elements.

Space defense regions illustrated in this book purposely are designed to be similar to traditional terrestrial AOR (Aera of Responsibility) concepts to enable better linkages between space warfighters and terrestrial operators."

Charles Williams, Rear Admiral (ret.) United States Navy,
Former: Assistant Secretary of the Navy, present President and
Board Member of the Navy League of the United States.

"Paul Szymanski and Jerry Drew in *The Battle Beyond: Fighting and Winning the Coming War in Space,* set the stage for practitioners of war in space and space warfare. From the details of symbology for space operations to the conception of space warfare into the larger enterprise of all-domain operations, the authors have created a seminal work that forms the basis for warfighting constructs that exploit the advantages of space. It provides critical and creative thought about each of the levels of war vis-à-vis the space domain and how they interact. Discussing the similarities as well as the differences between space and terrestrial operations, it addresses the elements of warfare in the context of how they might best be applied capitalizing on the warfighting domain of space. This is a must read—not only for space professionals—but for practitioners of military operations in all domains as winning in war today is wholly dependent on winning in space. This is a magnificent and very complete book!"
Lt Gen USAF (Ret.) David A. Deptula, current Dean of The Mitchell Institute for Aerospace Studies
former: Deputy Chief of Staff (DCS) for Air Force Intelligence, Surveillance and Reconnaissance; and, Director of Air And Space Operations, Pacific Air Forces

"When Gene Roddenberry's *Star Trek* first appeared on national television in 1966, viewers were captivated by the show's theme: 'Space... the final frontier!' Americans were the first on the moon, but it would take a half a century for America to recognize space as a warfare domain followed by the instantiation of the US Space Force. Space is now a contested area and we are playing a desperate game of catch up with our adversaries. Accordingly, Paul Szymanski and Jerry Drew's book could not have come at a better time. For those unfamiliar with intricacies of the "final frontier," they break it down into easily understandable prose and applies the warfighting principles of Sun Tzu, Napoleon and J.F.C. Fuller and others, in an operational design for space. They introduce us to the tools of space warfare from denial of service attacks to kinetic kill vehicles and grappler attacks, with an explanation of the broader impact of each. Furthermore,

Szymanski and Drew give us a healthy dose of the terminology and symbology used in space warfare. A close reading can make a neophyte into an overnight expert. Finally, they apply centuries of battlefield stratagems in creating an operational design for defensive and offensive operations in space, including a robust discussion of Rules of Engagement. A fascinating read and a must read primer for military and civilians alike who wonder what a conflict in space portends for our future."

James G. Foggo, Admiral USN (ret.)
Current Dean of the Center for Maritime Strategy, former: Commander Naval Forces Europe and Africa; Commander, NATO Allied Joint Force Command Naples; and, Commander, US 6th Fleet.

"The warfighter today is highly dependent on space-based capabilities for superiority on the battlefield. Unlike more conventional warfare, the battlefield and tactics in space are neither intuitive nor well understood by the vast majority of warfighters. In this book, the authors establish an actionable framework for understanding space warfare as well as a context for warfighter discussion and ultimately development of effective strategy. *The Battle Beyond* is a must read for all levels of command."

Marc Pelaez, Rear Admiral, United States Navy (ret.)
Former: Chief of Naval Research, Nuclear Submarine Commanding Officer, Vice President, Newport News Shipbuilding

---

"*The Battle Beyond: Fighting and Winning the Coming War in Space* by Paul Szymanski and Jerry Drew is nice work, and is pleasing to read. It's clear this book and I have similar experiences and views of the military art, not just the space domain. Finally, a comprehensive, classical military analysis of the new war-fighting domain of space that space professionals can read without fear of 'cringe-worthy' statements by well-meaning, military practitioners without space experience. In addition, the symbology was extremely helpful in showing the completeness and depth of your book."

Jim Armor, Major General, USAF (ret.)
Former: Director, National Security Space Office; Director, SIGINT Acquisition and Operations, National Reconnaissance Office; Director, NAVSTAR Global Positioning System; FAA Commercial Space Transportation Advisory Committee (COMSTAC); NASA Advisory Council (NAC); and AIA Space Council.

"To succeed in any conflict, control of the higher ground is essential. In current and future conflict, dominance of that higher ground is limited not to just the air domain, but depends on mastery of the space domain as well. In *The Battle Beyond,* Paul Szymanski and Jerry Drew do an incredible job of explaining what warfare in space will look like, using the foundational principles of Doctrine, Strategy and Tactics. They also outline the amazingly technical component of space mastery, requiring, at times, new ways of thinking.

Szymanski and Drew combine traditional theories of warfare from luminaries like Sun Tzu, Clausewitz and Fuller with clear explanations of the uniqueness of space warfare, while also stressing the importance of integration of space capabilities with those of air, land and sea. They lay out a roadmap for what needs to be done to give war-winning capabilities to decision makers. This book should be read by space warriors tasked with bringing space operational concepts to bear on the next fight, as well as warriors from the other domains seeking to fully understand what space can do for them. Space warfare is a difficult subject and we need to get this right in order to win. This book can help—A LOT!"

Maj Gen, USAF (Ret.) Gary R. Dylewski
Former: Director of Operations, Headquarters Air Force Space Command; Commander, Space Warfare Center; and Commander, Joint Task Force–Southwest Asia

"*The Battle Beyond: Fighting and Winning the Coming War* in Space provides policy makers and practitioners alike a framework on how to understand, operate and win in the complex domain of Space. Intellectually thought provoking, the authors weave relevant historical examples into today's realities faced by our national decision-makers regarding Space as a domain. More importantly, it provides meaningful approaches to how Space operations can best be integrated into all aspects of tactical, operational, and strategic support necessary to secure and to obtain national goals and objectives. Great job on this book. I look forward to seeing this work published and widely read!

Kurt Sonntag, Major General, US Army (ret.)
Former: Commanding General of the John F. Kennedy Special
Warfare Center and School; Commanding General, Combined
Joint Task Force – Horn of Africa; Commanding General, Special
Operations Command South; Deputy Commanding General
United States Army Special Operations Command.

"Paul's and Jerry's book could not be coming at a better time. Space warfare is critical to our success as a nation. We rely heavily on space for communications and command and control in everything from tactical engagements to nuclear command and control. Space is vital to position, navigation, and timing. In fact, the banking systems of the world rely on US and allied control of space to ensure that financial transactions are captured and validated. Space is key to our intelligence capabilities from visual to radio and human intelligence to open-source information. A former director of the US National Reconnaissance Office used to say, 'decades to build, seconds to destroy.' Our space assets are fragile, and we need to develop the space-based tactics, techniques, and procedures to ensure a credible deterrence to malfeasance by bad actors. Paul and Jerry have captured the essence of space-based defense, and I highly recommend this book for all, and serious national-strategic thinkers."

Richard C. Staats, Ph.D. Major General (ret.)
US Army, former: Commanding General of 316th Expeditionary Sustainment Command; Strategic Planner, The Joint Chiefs of Staff; and current Director for the MITRE Corporation.

"Space warfare theorists Paul Szymanski and Jerry Drew have made a substantial contribution to professional military education with their new book, *The Battle Beyond: Fighting and Winning the Coming War in Space*. Both prolific writers, Szymanski and Drew recognized the dawn of space as a critical warfighting domain long before most others and have dedicated years to thinking and promoting a national dialogue about it. This work, analyzing the ends, ways, and means associated with prosecuting a military campaign in space across the strategic, operational, and tactical levels of warfare, is arguably the pinnacle achievement of their cumulative study.

Szymanski and Drew approach space warfare as zealously as Mahan and Corbett explored naval warfare and Douhet and Mitchell examined air power. To their great credit, however, they do not view space as a warfare domain in isolation. Rather, they point out its important relationship with each of the other domains. Szymanski and Drew account for the consistent nature of war while concurrently highlighting space's impact on war's constantly evolving character. Indeed, military professionals will do well to add *The Battle Beyond* to their libraries alongside Sun Tzu's *The Art of War*, Jomini's *Principes de la Strategie*, and Clausewitz's *On War*. For our nation's new Space Force Guardians, this book should stimulate professional debate and growth. Those with a broader interest in national security who desire to gain an understanding of the vital role of space in current and future conflict will also find it a worthy investment."

BGen, USMC (Ret.) N. L. "Norm" Cooling
Former Assistant Deputy Commandant for Plans, Policies, & Operations at Headquarters, United States Marine Corps

"The modern world is not only highly connected but also highly contested. With so much of our economy relying on space-based capabilities, it is not surprising that space is now a contested domain—we must be prepared to fight and prevail. Paul and Jerry have written a compelling treatise on the battle in space—they've taken theory and concepts proven over centuries of warfare and translated their relevancy to warfare in the space domain. This seminal work will quickly become required reading at war colleges around the world."

James "Cliffy" Cluff, Brigadier General, United States Air Force (ret.)
Former: Military Assistant to the Under Secretary of Defense, Intelligence; Joint Staff J3, Deputy Director for Global ISR Operations (J32); and, Vice Commander, 25th Air Force.

"A generation of Americans have enjoyed the benefits of space-based systems. Not only are those systems vulnerable, but space itself is the primary battlefield; a battlefield often neglected until now. Paul Szymanski is America's premier space strategist. In his new book he highlights the systems at risk and delves into the tactics and strategies necessary to win the space war. Mr. Szymanski merges space systems and capabilities into theory and doctrine utilizing the timeless Principles of War. His masterpiece is a critical 'must-read' to understand, and dominate, the future battlefield!"

Air Force Brig. Gen. Talentino C. Angelosante
Former: Mobilization Assistant to the Commander, Air Force Recruiting Service, Joint Base San Antonio, Randolph Air Force Base, Texas; and Deputy, Current Operations, Operations Directorate, Camp H.M. Smith, Hawaii

---

"This is an outstanding book. This book brings into focus the importance of strategically engaging space militarily. The book is great at pointing out how the domain of space applies to the military and the complexities for civilian society. As a retired General Officer and now working in the civilian side space program, the book is good at showing how many entities have a hand in space and space warfare. An outstanding read, a must for space warfare strategy."

Brigadier General Ronald W. Wilson, US Air Force
Former: Director of the Joint Staff for the Michigan National Guard; Commander, 110th Attack Wing; participated in OPERATIONS SOUTHERN WATCH, ALLIED FORCE, IRAQI FREEDOM, and ENDURING FREEDOM.

"Paul Szymanski and Jerry Drew have written the ultimate guide to this new and complex domain of warfare. As they say in the book: 'the fundamental problem—the most essential ingredient for winning the coming war in space—is language.' And the language of space warfare since it was adopted as an operational domain of warfare by the NATO allies in 2019, is foreign to most military practitioners today. Paul and Jerry's text overcomes that obstacle in this book. They carefully and thoroughly explain the history and theories of space warfare for the space neophyte. This well-written treatise is a wonderful introduction to space warfare and a 'must-read' for practitioners to stay abreast of concepts developed in the study of warfare."

Brigadier General (ret.) Ronald S. Mangum, US Army, PhD, JD,
Former: Commanding General, Special Operations Command Korea; Commanding General, United Nations Special Operations Component.

"The space theater of operations is traditionally defined as comprising of outer space and celestial bodies, where, as foreseen by the relevant international law, States cannot claim or exert sovereign or exclusive rights in space. If space has not yet become like Earth, it has certainly become crowded, and commensurate with its relevance to our lives, space is more obviously than ever a domain of confrontation. Space war is yet to come, but shall come soon. And Land, Sea, Air, Cyber and Special Operations, be it for Intelligence, Communications, Meteorology, Ops Command and Control purposes, already draw heavily from the resources that Space operations offer. If the US and some other Nations have published space doctrines that draw adequate strategic conclusions from this state of affairs, they were not complete before the publication of Paul Szymanski and Jerry Drew's important new book, which is a single place to find a theorization of Space warfare, from the strategic to the tactical level. It is extremely well informed, built for practitioners and thought provoking. Remarkable, on all points."

Brigadier General (ret.) Xavier Périllat-Piratoine, French Air and Space Force
Former: Legal Adviser to the Deputy Chief of Staff,
International Relations at French Ministry of Defense; and,
NATO ISAF Theatre Financial Controller, Afghanistan.

"This is a captivating and inviting exploration of space warfare new doctrine supported by a comprehensive authors' detailed adaptation of classical concepts.

Like air power, the attributes of space power are related to the ability to exploit the vertical dimension, albeit to a significantly greater degree, as space can provide a truly global capability. The vast majority of military operations could not be sustained without space capabilities. Space power is inherently joint, because although it can be used to influence activities in space, it is primarily used to enable effects and exert influence in other domains, principally for communications, meteorological, positioning, navigation and timing functions, as well as for intelligence, surveillance and reconnaissance."

Rear Admiral (ret.) Gianluca Massucco, Italian Navy
Former: Strategic Development Staff Officer of NATO
CMRE (Center for Maritime Research and Experimentation);
and, Strategic Planning Division, Deputy Assistant Chief
of Staff for NATO Supreme Allied Command Europe.

"The reflections and ideas elaborated by Paul Szymanski and Jerry Drew are not only effective and rigorously scientific but also offer a complete overview and detailed analysis on what outer space warfare consist in and on all its specific aspects. Reading this book has been for me a unique opportunity to better understand the future of modern warfare, with a very concrete and practically useful approach. I recommend reading this book to all experts as well as persons simply interested to approach outer space warfare and related matters."

Rear Admiral Cesare Ciocca (ret.), Italian Navy
Former advisor to Italian Defence General Staff and the Italian Permanent Representation to the Political Security Committee of the EU, and NATO Defence College; current: Italian Atlantic Council and the VP of Atlantic Treaty Association.

"At first, I thought of Paul Szymanski as an unconventional personality, a visionary. Instead, I appreciated his clear analyzes, his genuine assessments, his original solutions and, in particular, his huge effort on updating the principles and the tools available to the Commanders at various levels to face, manage and dominate complex scenarios that are more relevant today than ever. With admirable dedication, Paul Szymanski and Jerry Drew have opened a new path and offered an essential guide to fully understand and approach the 'Outer Space' dimension of Warfare, urging the need to make them an imperative priority on Decision Makers' agenda for the years to come."

Antonino Inturri, Army Brig. Gen. (ret.)
Former Italian Ministry of Defence and Military Attaché for the Army at the Italian Embassy in Berlin.

"I read this book by Paul Szymanski and Jerry Drew, and really admired it for the new avenues of research and study it opens: for our Grand Strategy, Military Strategy, and war and warfare in general. As it points out, we need to establish the common language of interoperability and compatibility of all players and systems involved, whether terrestrial or space, in order to be able to reach a common dogma of operations and success.

This book will re-write military history, and become a foundational classic in military doctrine, due to its incorporation of all thoughts from philosophy to practice in this critical field. It will help considerably in our national security demands to be met in a timely manner, as we move along this world of increasing threats and crises. Congratulations for this unique great piece of excellence! It is my honor and privilege to endorse this book, and to offer my knowledge and expertise for further work on this critical subject."

Admiral Peter (Panagiotis) Kikareas Ph.D. (ret.)
Hellenic Navy (Greece), Current President of World Foundation For Peace and Security; Former Director of NATO Operational Planning for the Supreme Headquarters Allied Powers Europe (SHAPE).

———

"In my four decades of active service and another decade of studies on Military Science, I have not come across any book on warfare that has discussed the entire gamut as *The Battle Beyond: Fighting and Winning the Coming War* in Space by Paul and Jerry. Frankly, I have brainstormed my mind to see whether the author could have added even one additional aspect; failing to find any, I have given it up. This exhaustive manual on space warfare encompasses Grand strategy, military strategy, operational art, and right down to the tactical level.

If that is not enough, it also includes the technical aspect; without which it is impossible to comprehend its nuances and apply it coherently in multi-domain operations. It is replete with illustrations and considers several contingencies for better comprehension by citing live examples. The authors opine that the space war is not a future possibility but is waging right now and will form an integral part of the future multi-domain war. A brilliant book; a must for political leaders, military planners, bungling bureaucrats, and field commanders from generals to colonels. A superb book by Paul Szymanski and Jerry Drew. It is exhaustive and illuminating."

Lt Gen PG Kamath (ret.)
Former Commandant of the Army War College, Mhow, India.

———

"Right from the first satellite launched by the Soviet Union in 1957, space has moved from an isolated domain to one that integrates the operational capabilities of a nation in the traditional domains of land, air and sea. The use of space assets as an effective force multiplier has also been an important factor for countries to pursue their space programs with renewed vigor, both for offensive and defensive aims. With space assets becoming increasingly crucial to terrestrial military operations, countries have begun to view space as an exclusive warfighting domain.

A domain that for a long time remained the monopoly of two or three big powers, is today a crowded, congested and contested territory, with plethora of players including non-state actors. Space wars would continue to be waged in the near future due to the importance of this domain as the integrating domain for all the other traditional domains. It cannot be isolated from the conflicts raging in all the other domains as today's wars have become multi-domain. Space the Final Frontier remains far too important to the ultimate outcome of the terrestrial battlefield. Outer space once viewed as only fantasy may now be closer to reality than many realize.

Space has hitherto, largely been used as a domain to gain national prestige by nations being one up over their competitors. The strategic importance of space has become an imperative especially for countries which aim to assert military dominance over others. The exploitation of space has sparked a relentless race to achieve supremacy over others in this domain. Whilst nations may differ in their view of the strategic importance of space with some concluding that space will soon be a warfighting domain others view space as a medium to obtain an edge in terrestrial conflict through information superiority. However, there is little doubt that space as a domain will have a significant role in how terrestrial conflicts play out in the future. Paul Szymanski and Jerry Drew's latest book comprehensively encompasses the above viewpoint and brings forth the growing importance of Space Domain in the 'Competition Continuum' all across the Globe. The book makes interesting reading both for the first timers as well as the practitioners of Warfare as a profession."

Lt Gen Rajeev Sabherwal (ret.)
Former: Signal Officer–in–Chief, Indian Army; and, Commandant (Chancellor), Military College of Telecommunication Engineering.

"The relevance of space is substantive and growing, and space is key to all other military capabilities. As interstate competition heats up, an increasing number of states have launched satellites into orbit. Paul Szymanski and Jerry Drew's study provides relevant insights in the future of space competition and its role in military operations, including those on earth. Interesting read!"

Hans van der Louw, Major General, Royal Netherlands Army (ret.), former: Chief of Military Royal Household to His Majesty, King Willem–Alexander; Commander of 43 Gemechaniseerde (Mechanized) Brigade; Military and Assistant Defense Attaché to Royal Netherlands Embassy, Washington DC (US).

THE BATTLE BEYOND

www.amplifypublishing.com

For more information, please contact:
Amplify Publishing, an imprint of Amplify Publishing Group
620 Herndon Parkway, Suite 220
Herndon, VA 20170
info@amplifypublishing.com

Library of Congress Control Number: 2022908386

CPSIA Code: PRV0823A

ISBN-13: 978-1-63755-071-7

Printed in the United States

This book is dedicated to my wife, Darlene, who is the love of my life.

I also acknowledge former classical military strategists, such as Sun Tzu, Clausewitz, Jomini, Liddell Hart, Fuller, Corbett, and Mahan. May this book be a significant addition to the literature of warfare that has been written before and in the future for the new environment of military outer space warfare.

—P.S.

For Meagan, and

in grateful appreciation for the Space Warriors, who are always forward.

—J.D.

"Accepting observation in its everyday sense, it is needless for me to say much, for its utility is self-evident. Some people are very observant, others see next to nothing; some only see small things, others only big, and most only see what others see, and what others see is very often not worth seeing."

—J. F. C. Fuller, *The Foundations of the Science of War*

THE
BATTLE
BEYOND

Fighting and Winning
the Coming War in Space

PAUL SZYMANSKI & JERRY DREW

amplify

an imprint of Amplify Publishing Group

Contents

List of Figures

List of Tables

INTRODUCTION

War versus Warfare

He who exercises no forethought but makes light of his opponents is sure to be captured by them. Thus, what enables the wise sovereign and the good general to strike and conquer, and achieve things beyond the reach of ordinary men, is foreknowledge.
—Sun Tzu, *The Art of War*

Like so many elements of contemporary war, the word *war* itself comes into the western lexicon from the German. It is related to the Old High German words *werren*, meaning "to confuse," and *werra*, meaning "strife."* Whatever else war might be—the continuation of politics, hell, a "disease," an "act of murder," or a "defeat for humanity" †—confusing strife seems to be as good a definition as any. In the twenty-first century, war follows the decision of the state to employ available military means to achieve its policy objectives. It is an acknowledged hostility that involves various government agencies as well as the various military arms.

Warfare, in contrast, is the application of military means to the war effort. It is the practitioners' business—both of the generalists and specialists— and its practice includes knowledge and experience unique to the multitude of tribes. In land warfare, the tribes are the infantry, armor, artillery,

---

*   *Merriam-Webster,* s.v. "war," accessed March 12, 2021, https://www.merriamwebster.com/dictionary/war.

†   Clausewitz, Sherman, Saint-Exupery, Einstein, John Paul II

logisticians, and others. In maritime warfare, the tribes are surface warfare, subsurface warfare, and naval aviation—to name a few. Marines constitute their own tribe. In air warfare, the tribes include fighter pilots, bomber pilots, and transport pilots. In space warfare, the tribes are emerging. The list may include groups of experts in missile warning, electronic warfare, communications, intelligence, cyber warfare, and orbital warfare. Their skill sets will remain simultaneously unique to their domains and complementary to the skills of the other groups with whom they engage in joint and combined arms efforts.

As in land, maritime, or air warfare, the various disciplines of the space domain may be used independently, but the options available to friendly forces expand greatly when multiple disciplines coalesce into a common effort (i.e., in joint combined arms operations). In the same way, the effectiveness of the entire military increases when the means of multiple domains coalesce (conceptually, multidomain or all-domain operations). As governments strive toward their policy objectives, their own effectiveness will likely increase when national leadership aligns its various means from across the whole of government (doctrinally, this is the definition of unified action). It follows that if governments are not understanding of the space environment or not decisive in the employment of space forces, the effectiveness of military operations—and by extension, the attainment of policy objectives—may suffer.

Within the context of the space domain—a domain that includes not just orbital space but also the ground stations and radio signals operating to and through space—the terms *war* and *warfare* carry significant implications. If war involves all elements of national power, then nonmilitary entities that develop or employ space systems on the nation's behalf require consideration and depending on their function, may be subject to enemy action in a conflict. Organizations as diverse as the National Reconnaissance Office (NRO) and the National Oceanic and Atmospheric Administration (NOAA), along with commercial companies, academic institutions, and think tanks come into play. With this perspective in mind, it may be appropriate to ask (if only in private), "How is NASA contributing to the war effort?" Through human spaceflight, for example, NASA may provide a bridge by which to maintain relations throughout a period of international conflict. The question of how NASA is contributing to space warfare would be less interesting. True, space technologies developed for scientific applications may readily transition to military uses, but NASA is a civil government agency dedicated to space science and exploration not to warfare. Space warfare, then, is primarily a military problem and thus requires an application of military thought.

For the uninitiated, the words *military thought* may ring oxymoronic. After all, don't soldiers and sailors simply follow orders, leaving the brainwork to their betters? In the twenty-first century it is time to abandon this vestigial stereotype once and for all. The conflicts of the future will require the critical and creative thought of every person involved and the best-trained and most educated service members ever produced.

The ability to win the "battle beyond" will depend upon ideas that are only now beginning to emerge. Fortunately, a rich body of military thought— history, theory, and doctrine—exists as a font of knowledge for both the novice and the experienced practitioner, for the warfighter and the civil servant, for the concerned citizen and the casual observer. This corpus enjoys entrenched principles that can provide an abundance of inspiration, but a danger also lurks just beneath the surface: the tendency to ingrain ideas into the soldiery, no matter how useful for historical armies, is also likely to deter originality. In the quest for military effectiveness, the opportunity for misapplication of ideas is everywhere, either in applying the wrong idea to a given situation or in applying the right idea in the wrong way. A third pitfall lies in failing to apply an idea at all—perhaps for lack of ideas or for a lack of determination and resolve. The first two mistakes may be excusable; let us never allow the third mistake.

In this discussion of space warfare and how to succeed in the endeavor, then, ideas are the primary aspect to consider. Fortunately, the history of military thought is rife with them, and many ideas link readily to space warfare. We do run the risk of misapplication, but we also have an opportunity to apply old ideas in new and creative ways. For example, the doctrinal principles of war—derived largely from the military theory of British Colonel J. F. C. Fuller a century ago—remain useful. Similarly, the doctrinal elements of operational art mostly hearken back to the multidomain theorist (née the naval theorist) Sir Julian Corbett. The problem then is not that there is a lack of military thought to apply to space warfare. The problem is that, given a military establishment with a mandate for all-domain synchronicity—but one that is largely unfamiliar with space operations—a large gap exists in the translatability of space operations into terms understandable across the entirety of the institution. A related problem is the need for specific space warfare language when existing language will not do. In a word, the fundamental problem—the most essential ingredient for winning the coming war in space—is language. Soldiers, sailors, airmen, marines, and guardians may not know what questions to ask about space, and space experts (even if they are soldiers, sailors, airmen, marines, or guardians) may struggle to explain their answers in commonly understandable terms. In this effort, established military language is a more useful tool than trying to use the technical language of an esoteric discipline. Using already familiar military language, principles, and symbols therefore will help nonspace personnel coordinate better with space planners and prepare for the space conflicts of the near future.

If space is a warfighting domain, as US policy now states, then it requires a language—to the greatest extent possible—that is familiar to other warfighters. This language draws from doctrinal sources, which draw from historical and theoretical sources, and is as much a visual language as a verbal one. The planning and execution of operations—regardless of the domain—requires visualization, especially for the three-dimensional complexities of outer space. To visualize, one must both know the terms and be able to communicate them through graphic representation. Icons on a map are as valuable to space operations as they are to land operations; it is only the map that changes or the lens through which we choose to view it.

One may choose to approach space warfare through a variety of lenses—policy, technical, institutional, ethical, and warfighting, to name a few. While these lenses are certainly interrelated, the primary audience for this discussion is the military practitioner. As a part of war, the first portion of this work considers aspects of grand strategy and military strategy in their relation to space warfare. A consideration of tactics and the necessary language (both extant and proposed) follows. Tactics combine to achieve objectives.

If these objectives are strategic in nature, we may speak of operational art as the process that links the tactical to the strategic. Theorists from Jomini to Fuller may have called this approach "grand tactics," an intuitive term if one approaches war from the bottom. With a working lexicon of tactics, we may next consider how to combine tactical action to develop operational concepts or more holistic courses of action. Command and control in, from, and through space is a significant hurdle in this endeavor because—although there are analogues in other-domain operations—the space domain poses unique challenges that, like operations in the cyber domain, are outside the common experience of most military personnel. Only through a language that enables a discussion of means, ways, and ends do we have the tools to visually depict—and therefore to manage—space warfare as a part of war.

A thought process that begins with a discussion of strategy and works its way downward may arrive at unique insights due to the path chosen. By first addressing strategy and then addressing tactics and building upward, we strive to construct a system that is flexible enough to contribute to any type of warfare (large-scale combat, counterinsurgency, network-centric, or whatever new model that may emerge) while also supporting whatever strategic approaches serve the interests of a nation. If the tacticians, operational artists, and strategists of the future are creative, they may even be able to contribute to multiple types of warfare and support multiple strategic approaches simultaneously and with the same set of limited yet globally dispersed means.

This is the ultimate challenge of space warfare and what will be required for the fight ahead.

Vignette: Russian Navigation Satellite (GLONASS) and Ukraine

In February 2014, Russian paramilitary forces invaded the Crimea to "destabilize the situation and, if possible, convince the new Ukrainian government to accept a federalization scheme that would reduce their power nationwide and allow Russia to have substantial influence over individual regions."* The grand strategic aim of the Russians—to gain authoritative influence over a neighboring country's government—necessarily involved a subordinate military strategy that employed both overt and covert means and extended into the space domain. In mid-March, Russia accused Ukraine of jamming a Russian communications satellite—a disruption attack that Russia claimed was in violation of the International Telecommunications Union charter.† Around the same time, the North Atlantic Treaty Organization (NATO) headquarters in Brussels, Belgium, experienced multiple cyberattacks.‡ Two weeks later, on April 2, Russia experienced a systemwide failure of all twenty-four of its navigation satellites (GLONASS, the Russian analogue to the United States' Global Positioning System) for a period of thirteen hours—a possible denial attack.§ Eight more satellites failed on April 14 for approximately half an hour, once again calling into question the integrity of the constellation.¶

The exact times and sequences of the possible denial attack on April 2, as reported in open-source media, hint at more than just an accidental occurrence. The outages began after midnight Moscow time and followed a numerical sequence—first GLONASS 9, then 10, 11, 12, 13—suggesting that Russia's explanation of a technical system failure may not be the only possible one.** Indeed, the data suggests the possibility of a deliberate sequencing of tactical actions to achieve a strategic effect (an example of operational art), in this case, possibly messaging the Russian government that it was

---

* Michael Kofman et al. *Lessons from Russia's Operations in Crimea and Eastern Ukraine* (Santa Monica: RAND Corporation, 2017), xii.

† Bill Gertz, "Moscow Accuses Ukraine of Electronic Attack on Satellite," *Washington Free Beacon*, March 17, 2014, accessed September 19, 2020, http://freebeacon.com/nationalsecurity/moscow-accuses-ukraine-of-electronic-attack-on-satellite/.

‡ Ibid.

§ Staff Writers, "Satellite Navigation Failure Confirms Urgent Need for Backup," *GPS Daily*, April 8, 2014, accessed September 19, 2020, http://www.gpsdaily.com/reports/Satellite_Navigation_Failure_Confirms_Urgent_Need_for_Bac kup_999.html.

¶ "The System: GLONASS Fumbles Forward," *GPS World*, May 1, 2014, accessed September 19, 2020, http://gpsworld.com/the-system-glonass-fumbles-forward/.

** "GLONASS Suffers Temporary Systemwide Outage; Multi-GNSS Receiver Overcomes Problem (Updated)," *Inside GNSS*, April 3, 2014, accessed September 19, 2020, https://insidegnss.com/glonass-suffers-temporary-systemwide-outage-multi-gnss-receiverovercomes-problem-updated/.

being watched and that it could be subject to military as well as economic consequences for its actions in Ukraine. A line-of-sight analysis based on the known outages and the known orbital locations of the GLONASS satellites at the times of their outages suggests a possible location for this denial attack—Alice Springs, Australia.*

Figure 1 depicts the three orbital planes of the GLONASS constellation (in light blue and purple loops). On April 2, 2014, a ground station in the middle of Australia would have had visibility of and line-of-sight access to the affected GLONASS satellites. Interestingly, if the site is moved two to three degrees west or east from Alice Springs, horizon constraints prevent access to the proper GLONASS satellites at the proper outage times. If this was indeed a deliberate interference with GLONASS, these actions were not subtle, possibly reinforcing the theory of a deliberate message to the Russians. Had a belligerent wanted to attack the GLONASS constellation in a more clandestine manner, it could have taken advantage of the constellation's geometry and accessed the constellation from nearly anywhere in the world, even possibly from a ship at sea or from an airplane flying over a remote area. A more dispersed approach or an approach that did not attack in numerical sequence would have provided the advantage of making attribution nearly impossible.

---

* Paul Szymanski, "United States Loses First Global Space War to Russians" (Albuquerque: Space Strategies Center, 2014), 3.

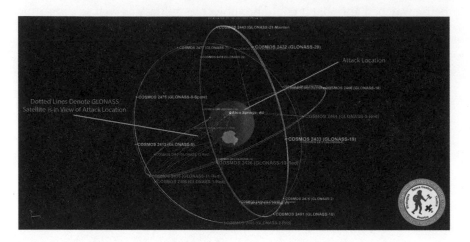

Figure 1: An Analysis of the 2014 GLONASS Outages. The Visual Shows that
Satellite Accesses were Possible from Australia. *Source:* Graphic created by
Paul Szymanski using the Satellite Orbit Analysis Program (SOAP) from The
Aerospace Corporation and orbital data from https://www.space-track.org.

Over the next six months, several additional events occurred that further
raise the specter of possible military action. In addition to their GLONASS
troubles, Russia's missile warning satellite COSMOS-2479 failed on orbit in
April, its $200-million communications satellite Express-AM4R exploded
atop a Proton-M launch vehicle crash on May 16, their communications
satellite Yamal-201 failed on June 6, and they temporarily lost control of
their research satellite Foton-M4 on July 19.* These events followed additional US economic sanctions and the threat/suggestion of Deputy Prime
Minister Dmitry Rogozin, head of Russia's defense industry, that "the United

---

*　"Russia Loses Its Last Early Warning Satellite," DefenseTalk, July 1, 2014, accessed September 19,
2020, http://www.defencetalk.com/russia-loses-its-last-early-warning-satellite-59992/#ixzz3B-
nZMVvuP; "Third-Stage Engine Glitch Causes Proton-M Accident," Space Travel, May 21, 2014,
accessed September 19, 2020, http://www.spacetravel.com/reports/Third_stage_engine_glitch_
causes_Proton_M_accident_999.html; "Failure Occurred in the Operation of the Russian Communications Satellite 'Yamal-201,'" RIA Novosti, June 6, 2014, accessed September 19, 2020, https://
translate.google.com/translate?sl=ru&tl=en&js=y&prev=_t&hl=en&ie=UTF8&u=http%3A%2F%-
2Fria.ru%2Fscience%2F20140606%2F1010895493.html; Abby Phillips, "Updated: There Is a Lizard
Sex Satellite Floating in Space and Russia No Longer Has It Under Control," *Washington Post*,
July 24, 2014, accessed September 19, 2020, http://www.washingtonpost.com/blogs/worldviews/
wp/2014/07/24/there-is-a-lizard-sex-satellitefloating-in-space-and-russia-no-longer-has-it-un-
der-control/.

States delivers its astronauts to the [International Space Station] with the help of a trampoline"—an example of the ties of space to diplomatic and economic aspects of grand strategy.*

While the GLONASS constellation, the COSMOS-2479, and the Yamal-201 all had obvious military uses, the Express-AM4R would have provided multiband communication coverage over the Eastern Ukraine (Figure 2).† To add to Russia's embarrassment, the loss of the Proton-M and its payload came just before Vladimir Putin visited Shanghai to sign a multibillion-dollar deal to provide natural gas to China—an issue that helped tip China's neutrality over Ukraine in Russia's favor.‡ The failure of two European Galileo navigation satellites to reach orbit (launched on August 22) may have been a response in-kind to the previous spring's attacks against GLONASS and to Europe's condemnation of Russian actions. It is worth noting that the failure of the Galileo satellites to reach orbit resulted from the failure of a Europeanized Soyuz rocket's upper stage.§

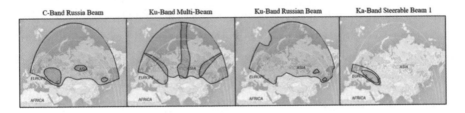

Figure 2: Satellite Communications Coverage that was Lost from the Destruction of *Express-AM4R*. *Source:* "Express AM4R," *SATBEAMS*, http://www.satbeams.com/satellites?id=2502

While it is possible that some or all of these satellite failures resulted from natural causes or human error, it is also possible that the string of Russian

*   "US Astronauts Should Use Trampolines to Get into Space, Russian Official Says," *Fox News*, April 30, 2014, accessed April 11, 2020, https://www.foxnews.com/science/us-astronautsshould-use-trampolines-to-get-into-space-russian-official-says.

†   "Express AM4R," *SATBEAMS*, accessed September 19, 2020, http://www.satbeams.com/satellites?id=2502.

‡   Jane Perlez, "China and Russia Reach 30-Year Gas Deal," *New York Times*, May 21, 2014, accessed September 19, 2020, https://www.nytimes.com/2014/05/22/world/asia/china-russiagas-deal.html.

§   Peter De Selding, "Galileo Launch, Initially Hailed as Success, Is a Failure," *SpaceNews*, August 23, 2014, accessed September 19, 2020, https://spacenews.com/41650galileo-launchinitially-hailed-as-success-is-a-failure/.

failures—or some part of it—was a deliberate attempt to degrade Russian military space capability, impose economic cost, and create room for diplomatic maneuvering. Furthermore, if some or all of these actions were deliberate, they follow a course beneficial for managing conflict escalation: temporary and reversible damage to the military GLONASS constellation, destruction of a missile warning satellite, and the loss of two communications satellites—the destruction cases being difficult to attribute to hostile action. None of these actions would have risked outright confrontation between Russia and the United States, but each had the potential to contribute to the grand strategy of the belligerents.

A discussion of which actions were part of a grand strategy to counter Russian activity in Ukraine remains, therefore, highly speculative. Regardless, the events of the spring and summer of 2014 provide useful examples of what *could* occur during a space conflict and are therefore useful as a springboard to further discussion on a host of topics. The brief discussion of Russian space losses during its actions in Ukraine highlights the intertwined nature of military activity with aspects of strategy, operational art, and tactics—all in relation to space warfare. These topics each require discussion in turn, but the first discussion must be one of strategy, the prime mover of the levels of war. To discuss strategy or the strategic level of war as it pertains to space operations first requires a baseline agreement of those terms. With deliberate explorations of both grand strategy and military strategy, it is then possible to explore their relationship to the other levels of war and their applicability to space warfare.

CHAPTER 1

The Language of the Strategic Level

Strategic Level of Warfare—the level of warfare at which a nation, often as a member of a group of nations, determines multinational (alliance or coalition) strategic security objectives and guidance, then develops and uses national resources to achieve those objectives.

Strategy—a prudent idea or set of ideas for employing the instruments of national power in a synchronized and integrated fashion to achieve theater, national, and/or multinational objectives.
—*Department of Defense Dictionary of Military Terms*

Section 1: An Overview of the Schema

Given its unique requirements, the military often co-opts words from other disciplines, changing a word's meaning to better approximate its military application. Other types of organizations do this as well, and in its various uses, *strategy* provides an example of this phenomenon, meaning different things to different audiences. In colloquial speech, for example, *strategy* has become a synonym for an approach, a plan, or even a technique (elementary school teachers, for example, often talk about strategies for controlling children's behavior). For the military, the term is even stickier, often blending parts of the colloquial usage with the approved definition (see epigraph) and a sense of something uniquely military. For this reason, and to set the conditions for a discussion about potential delineations of responsibility for both war and space warfare, a helpful distinction exists between the grand strategic (primarily concerned with the "war") and the military strategic (primarily concerned with "warfare").* Within the context of space warfare, a schema provides a road map for discussing the various aspects of the military strategic realm and highlights the importance of the institutional and the warfighting activities that the United States, particularly the US Space Force and US Space Command, must address to be prepared for the next war.

From the ancient Greeks onward, *strategy* enjoyed a specifically military meaning, originating from root words that suggested spreading out leadership or spreading out movement, only coming into nonmilitary use at the end of the nineteenth century.† The conceptual distinction between grand strategy and military strategy stems from the separation of the civil state and the military arm of the state—a problem that, as Sir Basil Liddell Hart noted, was less pronounced in Frederician Prussia or Napoleonic France because the monarch consolidated both grand and military strategy in his person.‡ In contemporary democracies, strategy proper has become the realm of the whole-of-government (in Liddell Hart's sense of "grand strategy"). Military strategy, in contrast, involves the interface of the military establishment with other players from the community of grand strategy (other government organizations and allies, for example) and with subordinate military elements. To put is slightly differently, grand strategy involves all means of

---

*   Joint doctrine uses the term *"theater strategic"* to refer to military activities happening at the highest levels of command within a theater. In the era of global conflict—and especially regarding military space operations—this term is potentially dangerous because conflicts cannot be isolated to a single theater. One may even go so far as to say that it would be irresponsible to perpetuate a system of military thinking that overemphasizes a theater-centric approach to strategy.

†   Online Etymology Dictionary, s.v. "strategy," accessed April 5, 2020, https://www.etymonline.com/word/strategy.

‡   B.H. Liddell Hart, *Strategy* (New York: Frederick A. Praeger, 1967), 333.

government; military strategy focuses on the employment of military means with the consideration of additional means, as available.*

Within the sphere of military space operations, military strategy interfaces with policy, with the other instruments of national power, and with operational-level warfare. Subsequent sections of this chapter deal with grand strategic considerations, and Chapter 4 deals specifically with the operational level of space warfare. The military strategic, then, serves as an interface between the grand strategic and the operational levels and exists within an operating environment that includes civil and commercial considerations, public opinion, and various threats to the state. In so interfacing, the military strategic must grapple internally with the tensions that exist between institutional drivers and warfighting needs. It must be simultaneously fighting and preparing for the next fight. The following model depicts a conceptual approach for how the operative aspects of the military strategic environment might interact with one another and shape operational-level warfighting (See Figure 3: A Schema for the Military Strategic).

---

* The authors acknowledge a certain amount of consternation from readers who feel very strongly that there need not be a distinction between grand strategy and military strategy—or even that there is no such thing as military strategy but only a sort of higher-level operational art. To best explain our points, we have settled on the military strategic, not as a distinct level, but as a critical interface between whole of government (grand strategic) and the employment of operational art. As the British statistician George E. P. Box said, "All models are wrong, but some are useful." The framework of the levels of war is itself imperfect, often requiring discursive footnotes to explain anomalies.

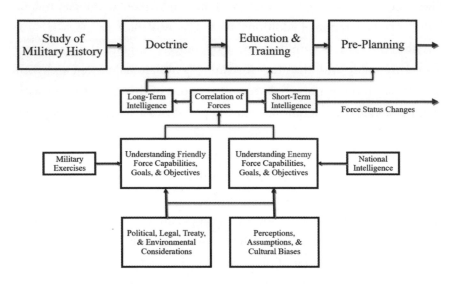

Figure 3: A Schema for the Military Strategic Interface. *Source:* Created by authors.

At the top left of the schema, the military strategic begins with the study of military history. Importantly for the space operations community, this study includes not only the study of military space operations (which trace back to the mid-twentieth century) but the study of all types of military operations. For the individual, exploring a variety of historical experiences provides professional insights into the application of one's own craft—regardless of one's specialty. For the institution, the understanding of military history, particularly of the subdiscipline of military space operations, takes on additional significance because military history is a prime informant in the creation of military doctrine, another important activity that occurs within the military strategic schema. Because of its significance, Section 5 of this chapter expounds upon the relationship between strategy and doctrine and provides a model for the creation of doctrine.

For the present, it is sufficient to say that doctrine provides a mechanism through which an institution develops a culture of warfare by establishing expectations, providing guidance, and creating norms of warfighting behavior. Because warfare, whether ancient or futuristic, involves human minds pitted against the minds of their enemies, doctrine, like the military philosophies of the past, becomes a tool of instruction both for classroom education (within the institutional domains of the various services) and for training (within the domain of warfighting application). A force that is educated and trained with the same doctrine will necessarily share many of the same ideas and will have a common frame of reference when the time comes for them to employ their craft.* In a word, they will share a language that, one hopes, will translate clearly to their joint force and multinational partners.

Like doctrine, multiple environmental factors—including logical, cultural, and emotional factors—influence education and training. Per the schema (bottom center), logical considerations include existing law, policy, and environmental considerations, not just in their application to the activities of the friendly nation but also in their application to the activities of adversarial nations, neutral nations, and nonstate actors. For the space domain as a physical location, the fragility of the environment to debris production stands out as another top concern. Certain military options—the use of an antisatellite missile to blow up a satellite, for example—may prove unpalatable if they are likely to cause negative diplomatic or economic repercussions (an interface with the other instruments of national power). On the more emotional side, strategy (both grand and military) must strive to account for perceptions, assumptions, and cultural biases of the relevant parties—a difficult task, even when considering one's own nation because it involves the amalgamation of the thoughts and beliefs of many individuals. In the present tense, the combination of logical and emotional factors manifests as a sort of running estimate for how the friendly force thinks the enemy will act and why. When there are radical cultural differences among nations, then there is an increased probability of political, diplomatic, and military misperceptions, which may lead to conflict escalation in space or elsewhere.

From these highest-level considerations flows a consideration of the military aspects of the problem. The strategist must seek to understand not only the political and cultural aspects of his enemy's activities but also their capabilities, goals, and anticipated objectives. Intelligence, the collection

---

*   Recall the warning from the introduction. It may not always be a good thing for everyone within the institution to have the same ideas.

and analysis of which is a shared burden across the whole of government, is essential for this purpose. In the preceding schema, "long-term intelligence" includes that intelligence that will inform the fight of the distant future—collections on an enemy's nascent antisatellite missile development program, perhaps. Oppositely, "short-term intelligence" includes the intelligence that is most relevant to the current fight. Here the dual nature of the military strategic manifests; it is both fighting and preparing to fight. On the friendly side, exercises, wargames, and simulations inform the institution of its own strengths and weaknesses, providing an intelligence of sorts about its own formations' locations, capabilities, and fighting status. All of these activities are essential in the continuous preparation for the next conflict.

Externalities—from the political to the cultural to the material—influence the activities of the military institution. When combined with the experience of the individual who has been educated and trained by the institution, the output of the military strategic is the early stages of planning for military conflict, here called "preplanning." This is deliberately a nonspecific term that involves setting the conditions in both the institutional and warfighting arenas. Within the bureaucratic structure of the US military system, some of the elements of the preceding schema occur within the institutional portion of the bureaucracy (i.e., the military services), and some occur within the warfighting portions (i.e., the combatant commands or joint task forces). Some activities, like training, occur within both arenas, and some activities, like doctrine development, are the responsibility of the institutional elements but require significant inputs from the warfighting elements. This schema, then, does not strive to clearly delineate responsibilities between various organizations whose structures and interactions vary widely. What it does do, however, is suggest that the activities of the military strategic interface, whether institutional or warfighting, bear directly on the operational level (see Chapter 4) and stem directly from the activities of policy and grand strategy.

Section 2: Policy and Grand Strategy*

The reestablishment of Space Command, the creation of a space force, and the designation of space as a warfighting domain provide recent examples of policy decisions about the use of space for military purposes. These policies signaled a clear willingness to expand US capabilities for space warfare and to engage more aggressively in that arena. Moving forward, future policy decisions must continue to address questions of political willingness to engage in space warfare, how best to engage in preconflict preparation, how to best control conflict escalation in space, and how to most effectively mitigate postconflict consequences (for example, space debris mitigation and the potential realignment of allies due to the diplomatic consequences of major space warfare events). Such policy decisions will shape the strategic environment and the strategic approaches that emerge, especially postconflict.

Conceptually, policy drives the need for a grand strategy to implement it, and grand strategy is the prime mover in the traditional hierarchy of the levels of war. In this hierarchical view, grand strategy drives military strategy, military strategy drives action at the operational level of war, and the operational level drives the tactical level. In reality, however, such a conception is too clean. Information-age militaries manifest as hybrid hierarchical-network systems, combining elements of top-down direction and information-based reaction. The levels of war—such as they are—interface on a continuous basis, informing, limiting, or expanding the available set of options in order, one hopes, to most effectively contribute to policy goals. When one level of warfare breathes out, another level—or perhaps multiple levels—breathes in. The recent coronavirus epidemic aptly illustrates this interconnection.

On April 5, 2020 (a Sunday), the secretary of defense signed a memorandum updating pandemic response guidance—one of many such updates throughout the crisis. By that evening, because of the ubiquity of smart

---

*   This section narrativizes many of the rules of space warfare developed by Paul Szymanski over more than fifty years of experience in the field. This list was originally published as the article "How to Fight and Win the Next Space War," *Air & Space Power Journal*, Air University, March 24, 2015.

phones and laptop computers, the policy had reached the First Space Battalion (an organization of roughly 430 soldiers), who faced the balancing act of implementing the new guidance immediately or following the standing guidance published by their higher headquarters. The military strategic had breathed out in order to preserve the force and to retain options for grand strategic action; all other levels breathed in the new information almost simultaneously.

In the hierarchical process characteristic of traditional armies, the Department of the Army would have received the Department of Defense's guidance, digested it, refined it for the service's peculiarities, and asked for clarification on unclear directives. Each echelon between the Department of the Army and the battalion would have followed similar procedures, taking at least a day or two at each echelon. Although chains of command vary widely within the army, in a representative case, the information may have flowed from the Department of the Army to Forces Command (a four-star headquarters) to a corps headquarters (a three-star one) to a division headquarters (two-star) to a brigade headquarters (a full colonel) and finally to a battalion headquarters (a lieutenant colonel).*

Under the rapidly changing situation of the coronavirus pandemic, however, a purely hierarchical information flow would have ensured the irrelevancy of the information by the time it reached the bottom. Given the circumstances, the challenges of interpreting unfiltered guidance probably caused less risk to the force than would have resulted from receiving outdated (but more complete) information on a hierarchical timeline. All echelons took in the Defense guidance and breathed out, responding with their own, sometimes disjointed, actions. The pattern of breath was erratic at first but soon resynchronized—partly because of the behaviors of the network, partly because of the steadying influence of the hierarchy as it gained traction—the vibrations largely settling before the next changes

---

* In the case of the First Space Battalion, the flow of information would have gone from the Department of the Army to Space and Missile Defense Command (a three-star Army Service Component Command) to the First Space Brigade (a full colonel) to the battalion. Because the battalion resided on Fort Carson at the time, information often arrived to the First Space Brigade from the Fourth Infantry Division, the senior army command on post.

occurred. In reacting this way, the whole force directly contributed to the grand strategic aim of preserving a fighting force for the nation's contingencies.

Because space warfare is coming of age in the twenty-first century, it is necessarily coming of age within a hybrid hierarchical-network structure. Indeed, space capabilities like satellite communications have allowed for the gradual transformation of militaries away from more strict versions of hierarchies; metaphorically, it is no longer possible for Frederick to breathe for the army. To be sure, the hierarchical aspects of the military still serve their purposes, but bureaucratic action, however necessary, cannot keep pace with information flow. It is therefore incumbent upon space operations forces to work under a system of higher leaders' intent with a clear understanding of what authorities are delegated to them (and what are not) and of what lies within the realm of strategy and what lies within the realm of tactics (a common point of confusion), especially when discrete tactical actions—the destruction of a missile warning satellite, for example—may have strategic consequences.*

As with all types of warfare, space warfare and the fieldcraft employed therein necessarily derive from policy, and the strategic approaches—both grand strategic and military strategic—must meet those policy aims. Due to the relative newness of space warfare, however, it may be difficult to match a policy aim to the appropriate military action with certainty. Based on a shortage of historical examples, an in-depth understanding of the diplomatic, informational, military, and economic ramifications of employing space weapons systems may not be possible. Is the destruction of a missile warning satellite, for example, a "shot across the bow" to message resolve, or is it an attempt to reduce surveillance coverage as a prelude to a nuclear

---

* For the military practitioner, the action of destruction is a tactical one; for the strategic authority, the decision is a strategic one. Similarly, the movement of a "strategic" asset on orbit is a tactical action that may play into a larger strategy, but the military force is not operating at the strategic level. These distinctions may sound pedantic, but they are necessary for clarity within space warfare, particularly when real-time responses may be required much faster than the highest headquarters can respond to them.

strike?* Within the context of the Cold War, the acceptable answer may have been more obvious, but in a multipolar world with a host of space actors, such considerations are highly contextual and vary depending upon the views of the belligerents. When the Russian COSMOS-2479 failed on orbit in April 2014, they did not automatically launch their Satan missiles. Ambiguity, then, over how an actor might respond may have significant deterrent value (though this idea has some danger in it because potential adversaries may have entirely different cultures with differing interpretations of the meaning of military actions). However, if an actor was willing to take advantage of an enemy's indecision, the secretive nature of space activities may provide opportunities to employ forces toward aims that lie outside the bounds of publicly stated policy.

In a related twist, policy makers and strategists alike may not fully understand the value of space to themselves and to their adversaries. In underestimating this value, they risk underresponding to adversary activity that may threaten access to space or spacecraft. In overestimating, they risk alienating allies or neutrals who may not approve of a response that they view as disproportionate. What is the appropriate response if an enemy destroys one of your missile warning satellites? In fact, responses from the enemy, from allies, from neutral actors, and from the population itself may vary widely, depending on a variety of circumstances. Space warfighters, therefore, must always pay attention to how their actions can impact the postconflict environment. If the space war is perceived as too severe in its implementation, negative consequences like allied realignments or political backlash are real possibilities.

Whatever the severity of the conflict, it is possible that military actions against space systems will not directly produce casualties, making orbital or ground-based counterspace actions seemingly innocuous (an underestimation). If one subscribes to this line of thinking, commanders may be tempted to be particularly decisive and coldhearted in their planning

---

*   It is overly simplistic to assume that the military view will be the more aggressive one. It is true that a military strategist may favor one course that would be unthinkable to a grand strategist, but it is equally likely that the grand strategist will envision a response that would be unthinkable to a military strategist. It was, after all, Bismarck who ordered the shelling of Paris, not Moltke.

and execution of operations. As Lieutenant General (Deceased) Roger DeKok, a former vice commander of Air Force Space Command, once stated, "Satellites have no mothers." Destroying satellites or interfering with their radio frequencies, in this view, takes on no more moral gravity than destroying an unmanned aerial vehicle or a radio tower and requires just as little courage. Further contributing to the sense of nonconflict are the facts that spacecraft are far away and hard to directly observe, actions against them may be difficult to attribute, and the consequences of the loss of a military satellite may not be discovered by the public.

In contrast to this view, it is also possible that escalation of military action in space will fire the imagination of the body politic—for good or for ill. The world may react viscerally upon learning that combat action has extended into space, especially if second- and third-order effects—like the wanton creation of space debris—increase risk to other satellites or to military forces in other domains. Indeed, the creation of a debris cascade may well make losers of all belligerents—and even of nonbelligerents—in the long term and of other-domain forces in the near term. This view, however, assumes that an adversary is dependent on space systems. An adversary that is not dependent on space systems (North Korea, for example) for its military or economic well-being may have an entirely different definition of winning. Such a nation might be willing to risk debris creation or even a nuclear detonation in space, if it imposes significant cost on its enemy and ensures the attainment of its policy objectives.

The diversity of worldviews and the potential for very different responses may seem undesirable. After all, how do we know how to behave if we have no agreed-upon standards of behavior or mutually acknowledged responses? For all of its potential for misinterpretation, this ambiguity also opens an interesting strategic potential: because there is not necessarily a common definition of winning, both sides may perceive that they have won—or at least not lost unacceptably. Because other actors likewise do not all share common views, it may even be possible for each side to garner enough support to justify its actions. And if both sides are content with the political outcome, further military conflict may be redundant—a not undesirable outcome. Just such a case occurred in the Iranian missile attack on US forces in January 2020. Iran was able to save face and gain regional prestige by responding to the US airstrike that killed Major General Qasem Soleimani, and the United States was able to downplay the attacks and emphasize Iran's military ineffectiveness. In these cases, it seems highly likely that the belligerents—and those observing the conflict—will derive quite different conclusions from the experience. These conclusions may dominate their strategic thought for decades to come, influencing the way they evaluate their enemies, evaluate themselves, and prepare for the next bout of conflict.

Whatever the outcome of the conflict, in a postconflict environment, all belligerents are likely to reassess their arsenals, building additional space and counterspace capabilities and possibly finding creative ways to reduce their dependencies on space systems. Weapons development and deployment will be covert, leveraging state-of-the-art technologies and driving proportionate efforts to increase intelligence-gathering efforts. Despite such efforts, the "fog of war" will remain very real in space warfare. Not only is it difficult to know what an adversary is thinking or what a nation is doing, but from a purely physical perspective, it is difficult to assess what is happening within the vastness of space and the dynamism of the electromagnetic environment. While operating unobserved provides certain advantages to any party capable of doing so, subtle messaging that allows for gradual conflict escalation requires the adversary to see your activity and assess its meaning. In this way, space operations set the conditions for additional "rungs" on the conflict escalation ladder. These gradual escalations, potentially hidden from public or international observation, provide an opportunity for relatively private direct messaging between belligerents. If the result of deliberate military action, the GLONASS outages from the vignette may have been such an attempt at gradual escalation. For now, these deliberate messages are special cases that may exist among space powers with sufficient technology to orchestrate such tactics. In general, and until space domain awareness capabilities achieve substantial advances, covertness and its offspring, surprise, will remain among the most valuable assets of space warfare.

Perhaps, then, the most important assets for senior military and political leaders will prove to be surveillance sensors with corresponding assessment algorithms to help make sense of the activities of tens of thousands of orbiting objects and hundreds of thousands of different satellite signals. The proper combination, location, and type of sensors and algorithms is a question of grand strategy because it involves research and development, procurement, forward basing, and considerations of overall architecture—in addition to the implementation of said systems. A healthy industrial base will remain a necessity, and the uncertainty over which technologies will prove most effective or what new technologies will emerge will require the institution and warfighters alike to retain optionality. No matter how robust the sensor network or its algorithms, however, because of the cleverness of human beings, especially under combat conditions, it will be impossible to reduce all vulnerabilities to space systems through a perfect knowledge of what is happening on orbit, within the electromagnetic spectrum, or even on the ground. For this reason, an overconfidence in one's intelligence-gathering may be particularly dangerous, leading to misunderstanding among decision-makers and an underappreciation of existing vulnerabilities. Whatever these vulnerabilities may be, adversaries surely will develop techniques for exploiting them.

As the need for space domain awareness demonstrates, an opportunity exists for the belligerent most able to make sense of the implications of space warfare to gain a relative advantage. As necessary as technical solutions are in this endeavor, however, it would be a mistake to focus only on technical solutions. Indeed, nontechnical solutions are more abundant and perhaps easier (and cheaper) to implement. What theoretical concepts, doctrine definitions, strategic approaches, or tactical tasks are optimal for space warfare remain open questions—ones that we will endeavor to address—and will most certainly vary depending on particular circumstances, but these are among the most important questions to ask. Because all information, knowledge, and understanding are imperfect, all parties will make mistakes in their assumptions and in their interpretation of facts. As in all warfare, small advantages may make large differences.

The desire to reduce the uncertainty of what could happen in a conflict or afterward can only be achieved through preparations. Whether we call these postconflict preparations or preconflict preparations remains a matter of perspective. Indeed, the United States, despite some continued military action, sits in an interwar period, and only time will tell what the outcomes of this interwar period will be. The designation of space as a warfighting domain provides an inflection point within this period, and as a result, we choose to consider the rest of the discussion in terms of preconflict preparations.

To set the conditions for the nation's progress moving forward—to put it in the position of greatest advantage—a grand strategy may involve treaty negotiations, allied partnering, and development of the industrial base. In these efforts, the space-related components of such strategies should address government agencies like NASA, NOAA, the NRO, the National Security Agency (NSA), the State Department, the Department of Commerce, and the military departments. Within the joint force (and closer to the realm of military strategy), Space Command will certainly be a significant actor, but all joint force commands require attention. These joint force commands will necessarily depend on Space Command to conduct activities on their behalf, but the priorities of Space Command will not always align with those of other commanders, forcing discussions of how best to integrate plans, synchronize operations, and reach a level of risk that is acceptable to all parties. It is likely, too, that Space Command will depend on other commands. During military action on orbit, for example, are there actions that commanders could take in South America or Central Asia that could increase the chances of success in space? Can Space Command deny, disrupt, degrade, or destroy terrestrial ground systems that are supporting satellites essential to special forces raids, air force attacks, or naval patrols in the Pacific? As a question of grand strategy, are there actions the whole of government can take prior to conflict to set the conditions for such options?

In addition to affecting domestic activities, decisions made in the preconflict environment will drive conclusions among the international community. Even without the creation of a cascading debris field that destroys billions of dollars' worth of orbital assets, the economic ramifications of military actions in space (or the fear thereof) could be significant. The loss of the Global Positioning System (GPS), for example, would manifest itself in the delayed transportation and shipment of goods and in confusion within the banking system, which uses GPS as a time source. The ensuing inefficiencies would slow commerce, reduce profit margins, and affect unemployment rates. Following such a crisis, national and international protocols, treaties, and rules of conduct would likely be ripe for change—or at least for discussion, and it is the duty of the strategist to take into account these potentialities before and during the conflict. Given the uncertainty of how a conflict in space may play out and of the effectiveness of such conflicts in achieving intended aims, preconflict preparations, particularly as they involve partners, take on new significance.

One way of shaping the preconflict environment is through international agreements, either bilateral or multilateral. International treaties with multiple signatories provide the most formal framework for a rules-based order, but there are many other less formal types of agreements available to diplomats. In this consideration may lie the most useful role of allies in space warfare—to shore up a nation's military strength through networked cooperation.* It is true that there are a limited number of countries with robust military space programs. Furthermore, due to the attendant need for covertness along with political sensitivities, sharing knowledge of one's capabilities, even with the closest of allies, can prove a challenge. It should also be noted that, though having allies with a shared purpose is generally good, this concept might not be as applicable to space as it is to terrestrial operations. For example, if the United States was at war in the Western Pacific, NATO satellites or antisatellite systems covering Europe would not be of much use. Even low earth orbit satellites that might "accidently" cover the Pacific theater would not have ground control and data reception assets in theater. Where NATO could assist, however, might be in taking on a larger role in Europe, thereby allowing the United States to focus more of its resources in the Pacific.

Even with a well-established alliance like NATO, planning military operations—even for a mission as routine as satellite communications—is difficult unless there are clear agreements in place. It is quite possible to share, trade,

---

*   James C. Moltz, "Coalitions in Space: Where Networks Are Power," *Space and Defense* 5, no. 1 (Summer 2011): 5.

or sell services like satellite imagery, communications bandwidth, or missile warning coverage. The trades may not even come in kind. For example, missile warning coverage may be worth a certain amount of airspace, a concession in a status-of-forces agreement, or host-nation logistical support. But in the grand strategic sense, these negotiations involve multiple government actors and take significant time and effort. In crises, such barriers may be lessened and processes accelerated, but if a conflict were to erupt tomorrow, belligerents will be limited by the capabilities available to them, the surest of those capabilities being the ones they control themselves. In open conflict, sharing or sales agreements—even longstanding treaties—may be suspended or even ignored. Even in peacetime, it may prove relatively easy to ignore treaties in space because of the vast distances involved, a general shortage of space surveillance equipment, and the infeasibility of adequately inspecting the interiors of every spacecraft. Compounding the problem, it is difficult to understand ownership of many mysterious space objects: as of April 1, 2021, the open-source space catalog lists 486 space objects that cannot be traced back to specific launches or countries.

As in the preceding discussion of ambiguous ramifications, the lack of a shared value assessment complicates both developing a strategic approach (for example, in navigating interactions with allies) and developing tactical responses (for example, developing maneuver and fires plans). Nonetheless, the employment of space forces in competition or in conflict—as with forces in any other domain—benefits from a political will focused on well-defined aims. In practice, those aims will likely shift throughout the conflict as the situation evolves, but having space forces that are superior in numbers and technology—if those are indeed metrics of a successful space force—may be useless if there is not the political will to employ them or if political will is divorced from an appreciation for the unique constraints of space warfare. To do this effectively, policymakers must understand the means available, the ways in which those means may be employed to achieve the desired aims, and the implications of such actions—in a word, the building blocks of strategy.

Section 3: Ends, Ways, and Means*

What emerges from the preceding assertions are a handful of the most significant aims of space warfare—the ends. Within this context, we are not referring to the ends of a particular campaign or war, which are necessarily case specific, but of the general aims of military space operations, including the service's and the warfighters' responsibilities, as they exist within the larger context of international competition. As long as competition remains, the "ends" may more correctly be labeled as conditions of a desired future state; they may not end, per se, but may continue to evolve as the operating environment evolves.

What are the ends, then, of space as a warfighting domain? Depending on policy mandates, a given adversarial context, or the vision of a commanding officer, these may vary considerably. However, in general (and resulting from the preceding exploration of policy and grand strategy), the desired future conditions necessary to prepare for the coming war in space fall under three categories: widespread understanding of the space warfare as a discipline, the ability to prevent the enemy from achieving surprise in the space domain, and the ability to posture for relative advantage against an enemy. Figure 4: Desired Ends of Space as a Warfighting Domain depicts this visually.

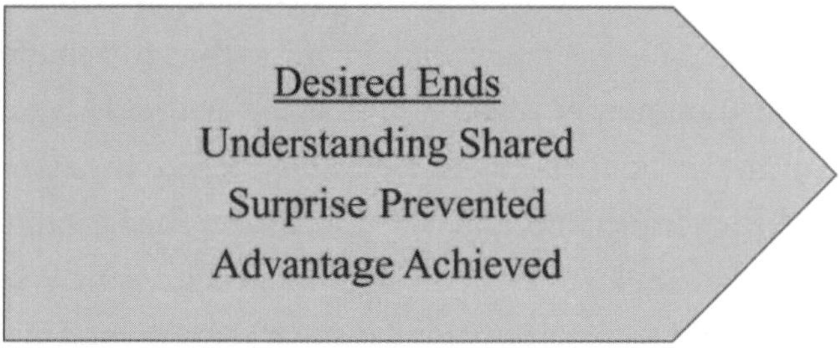

Desired Ends
Understanding Shared
Surprise Prevented
Advantage Achieved

Figure 4: Desired Ends of Space as a Warfighting Domain. *Source:* Created by authors.

---

*   The "Ends, Ways, Means" framework establishes the joint doctrinal method for strategy development. See US Joint Staff, *Joint Publication* (JP) 5-0, *Joint Planning* (Washington DC: Government Printing Office, 2017).

These proposed ends may be equally useful as grand strategic ends, as military strategic ends for a joint force commander, or as military strategic ends for a military institution, such as the space force. Given the means available to the United States in the twenty-first century, no one strategist could possibly understand how to employ all available means within a grand strategy—or within a military strategy, for that matter, but that should not stop us from trying our best. In one common framework, the instruments of national power are listed as diplomatic, information, military, and economic (DIME), but even these are vast categories. Despite their vastness, it is not possible to say that the inclusion of all of these instruments in one's strategy makes the strategy complete. Other instruments may be required, or some instruments may not even be necessary to achieve a given end. As a result of the number of different subdisciplines within the whole of government, specialization of individuals is required; grand strategy depends upon the expertise of constituent subcultures to contribute to its success. One of these subcultures is the space warfare subculture, and while space warfare may not always be required to achieve given strategic ends, the vignette preceding this chapter hints at the ways in which space warfare could contribute to a grand strategy.

That some strategies do not require all instruments should not be a surprise. Even in tactical plans, one may not always need tanks or fighter aircraft to accomplish one's objectives. Similarly, a national government may not always require a military element as part of its effort to achieve a policy aim. Such a situation may be difficult to imagine, particularly for military professionals in the current American system, where the size, funding, planning expertise, leadership, and risk tolerance of the US military make it a flexible tool for employment in most contingencies—the West African Ebola response of 2014, domestic natural disaster recovery, and maintenance of critical infrastructure provide just a few examples of how the military contributes to national aims outside of conflict. Military strategy is a part of this specialization, and in the same vein, one may, if one was so inclined, speak of diplomatic strategy, information strategy, or economic strategy.

For the twenty-first century, military strategy implies a comprehensiveness that touches all available military means (in the sense of large-scale operational art). What distinguishes the military strategic from large-scale operational art, however, is its direct interface with elements of national power. Military strategy reaches into available national means—to which the commander may have limited access or control—to contribute to grand strategic aims. It also may reach into the means of other nations through peaceful initiatives or military action. The military-strategic, then, is not a level of war—a term that implies an artificial separation—but, as stated before, it is an interface between the grand strategic and the operational level.

To put it differently, if grand strategy is primarily concerned with the components of DIME, then military strategy is concerned with the military aspects of those same elements. Figure 5 shows a simplified view of the elements of grand strategy, the military element (M) being only one. The military instrument often interfaces with the other instruments directly (with diplomatic representatives from the State Department for foreign military sales, for example), but within its own sphere, the military has diplomatic, informational, and economic elements (denoted in the graphic as MD, MI, and ME, respectively).* In this construct, the military does not orchestrate diplomacy proper (although it is very concerned with the outcomes of diplomatic activities), but it does interface with allies and partners in a military version of diplomacy. Such activities may be as simple as training with foreign militaries or as formal as participation in international councils. Similarly, the military is not responsible for the national economy or for the information that the nation puts out, but military action—whether domestic or international—affects local economic conditions, and information operations (ranging from public affairs to military deception) are integral parts of military planning.

---

*   The authors attribute this observation to Dr. Bruce Stanley, Professor of Security Studies at the School of Advanced Military Studies, Fort Leavenworth.

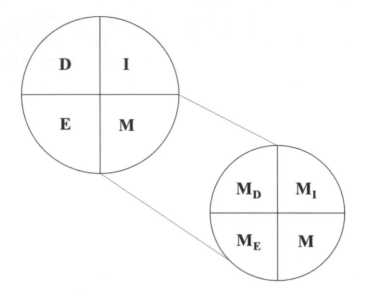

Figure 5: *Source:* Graphic created by authors based on an insight from Dr. Bruce Stanley.

Take as an example the Supreme Allied Commander-Europe (SACEUR) of the North Atlantic Treaty Organization (NATO). As the textbook example of an alliance, NATO is external to both the US military and to the US government as a whole, yet the SACEUR wears the additional hat of Commander, US European Command (one of the US military's combatant commands). As a combatant commander, it is debatable (from the standpoint of current doctrine) whether he resides at the operational or strategic level of warfare. In one interpretation, the strategic level of war starts with the chairman of the Joint Staff and proceeds upward (in the sense of "grand strategic"). Per the joint doctrinal definition, however, one might argue that since combatant commanders are heavily involved in determining "multinational (alliance or coalition) strategic security objectives and guidance" and obtaining the necessary resources, those military commanders are operating at the strategic level of war. The SACEUR's authority, of course, does not supersede that of cabinet secretaries, presidents, or prime ministers (who without question reside within the strategic level of war), but his expertise is certainly valuable in determining strategic approaches with allied nations through military and diplomatic interaction. Furthermore, as the interface between the operational level of warfare and grand strategic activity, the SACEUR serves as a notable example of the military strategic.

Within the initial space force, the chief of space operations stood at a similar interface between December 2019 and August 2020. During that time, General John Raymond oversaw the nascent institutional aspects of the space force—the administrative organizing, training, and equipping functions common to the services. In this role, he worked for the secretary of the air force and interfaced with large portions of government, including Congress and various federal departments and agencies—the grand strategic actors who would be involved in waging a war. As the combatant commander of US Space Command, he simultaneously headed the warfighting aspects of space operations, straddling the military strategic and the operational level and focusing on warfare. In this role, he answered to the secretary of defense. With the confirmation of General James Dickinson to succeed Raymond at Space Command in August 2020, the service and warfighting responsibilities were once again separate.

In theory, the potential benefit of combining the service chief with the field commander could have led to synergy between institutional and warfighting aspects of the military strategic. The historical roles of Ulysses S. Grant as general of the Army of the United States and of Helmuth von Moltke (both Elder and Younger) provide examples of this phenomenon, but in the twenty-first century the dual-hatting of General Raymond was only a stop-gap measure, the overly extensive scope of responsibility being divided as the service and the combatant command matured. To be sure, the acquisition and fielding of capabilities to meet warfighting demands remains a pressing need—arguably the reason for which the space force was created—but enough work remains in the designation and fulfillment of capability gaps to separate it from the warfighting.

The distinction between the warfighting aspect of military strategy and the institutional aspect of military strategy is the difference between a means-focused approach to building a strategy and a ways-focused approach to building a strategy. The first is necessarily the more practical, the second the more visionary, but both inform each other. To achieve a series of end-state goals, both approaches are necessary.

The Warfighter's Approach to Military Strategy: A Means–Focused Approach

Imagine you are a combatant commander in a wartime setting. You and your staff have identified a series of military objectives that you believe will substantially contribute to the policy aims of the conflict. To accomplish these objectives, you have a series of means at hand, and your staff has developed several approaches for the employment of these means (the ways). In this case, the ways depend largely upon what means are available at the moment. Apart from a handful of emergent weapons or tactics that are nearly ready for operational use, it would, after all, be foolish for a commander to fantasize about what means she wished were available. Similarly, there may be creative new ways in which to employ these forces, but employing formations in ways for which they have not been trained incurs a certain amount of risk. In the moment, then, the strategic approach depends largely on the means available and the established ways of employing those means. Creativity is bounded. To attain more or different means or to employ them in significantly different ways requires support from military and nonmilitary institutions: the Department of the Air Force may provide more aircraft and more satel-lites, the Department of State may engage in foreign military sales and other diplomacy, and the US Agency for International Development may provide humanitarian expertise—to name a few examples. The military services will work to provide improved forces to the fighting commander, and external departments and organizations may contribute to the goals of the military commander as part of the larger grand-strategic plan. The military com-mander's plan, when it includes instruments of grand-strategic power, exists at the interface of the grand strategic and operational levels of war—that is to say, within the realm of the military strategic.

Given the ends shown in Figure 4, the driving consideration for the warfighting aspect of military strategy—that is, for the near-term fight—is what means are available. The task of the military strategist then becomes linking the available means to the desired ends—that is, determining the ways or, one might say, developing the approach. Figure 6 depicts this means-focused approach to developing strategy, leaving the undefined ways as the most significant hurdle to overcome. A deeper understanding of existing means and the ways in which they have been employed in the past will allow for greater creativity in their future employment. Educating and training individuals to attain this depth rests with the institutional aspect of the force and is enhanced through operational experience.

Figure 6: The Warfighter's Approach to Military Strategy—a Means-Focused Approach. *Source:* Created by authors.

The Institution's Approach to Military Strategy: A Ways–Focused Approach

In the US military system, combatant commanders (or other joint force commanders) are responsible for fighting. The military services organize, train, and equip forces for fighting. These three functions may be said to be the generic ways of the institution. Their operative question is not "What means are available for fighting?" but "How will the forces we provide fight in the future?" Because institutional change and acquisition processes involve long lead times, these questions are largely speculative. Enter several of the portions of the preceding schema for the military strategic: military history, cultural factors, long-term intelligence, to name a few. Through studies, historical assessments, experiments, simulations, and wargames, the institution will determine how it envisions fighting in the future, and it will attempt to modify the means to allow for these new approaches.

Modifying the process used to produce Figure 6, we may now consider our ends and develop the ways to achieve them, not necessarily bounded by existing means. Figure 7 lists a few suggested ways. Each suggestion requires its own critical analysis. One might rightly ask, "How do we integrate existing capabilities?" If we are discussing the integration of space domain awareness sensors, for example, we may identify the need for more radars, for optical sensors in different locations, for more on-orbit assets, or for specialized interfaces between existing sensors so that they can share data. These designated needs become the basis for procuring new means, a function of the military services. If the warfighting commander designates such a need, he may request it through a variety of mechanisms, an operational needs statement being one of the most common.*

Suggested Ways	Desired Ends
Integrate existing capabilities.	Understanding Shared
Develop new capabilities.	Surprise Prevented
Educate & train the force.	Advantage Achieved

Figure 7: The Institution's Approach to Military Strategy—a Ways-Focused Approach. *Source:* Created by authors.

Happily, in June 2020, during the period when this manuscript was in draft, the Department of Defense published the *Defense Space Strategy Summary*, which corroborates the preceding theoretical assertions. The institution took a ways-focused approach to developing a strategy. After outlining the desired conditions (i.e., the ends) and the strategic context, the document lays out a strategic approach, concluding the document in a mere eighteen pages. Organizing by lines of effort (LOEs), the strategic approach includes the following:

LOE 1: Build a comprehensive military advantage in space.

LOE 2: Integrate military space power into national, joint, and combined operations.

LOE 3: Shape the strategic environment.

LOE 4: Cooperate with allies, partners, industry, and other US government departments and agencies.†

---

*   The chairman's Annual Joint Assessment provides another opportunity to designate capability gaps. In this effort, the joint staff compiles the assessments of the combatant commands and services to prioritize needs across the joint force.

†   US Department of Defense, *Defense Space Strategy Summary* (Washington DC, 2020): 6, https://media.defense.gov/2020/Jun/17/2002317391/-1/-1/1/2020_DEFENSE_SPACE_STRATEGY_SUMMARY.PDF.

Each LOE includes a number of objectives, further converting a highly conceptual approach into a series of tangible objectives toward which the force can focus its efforts. LOE 1, for example, includes the following objectives:

- Build out the US Space Force.
- Develop and document doctrinal foundations of military space power.
- Develop and expand space warfighting expertise and culture.
- Field assured space capabilities.
- Develop and field capabilities that counter hostile use of space.
- Improve intelligence and command and control (C2) capabilities that enable military advantage in the space domain.*

The *Defense Space Strategy Summary* of June 2020 provides a concise, ways-focused strategy to guide the institution, and all practitioners of space warfare should read it. Furthermore, its publication provides the opportunity to highlight two very important points about the development of a strategic approach. First, no two strategists or groups of strategists will devise the same ends or the same approaches to achieve those ends. The ends (or desired future conditions) may come from policy or grand-strategic guidance, but it may also be necessary to develop additional, complementary ends. The ways are very much open to interpretation and creative thought. The ways-focused approached outlined in the *Summary* did not include a graphic depiction—a missed opportunity—but based on the words, the graphic may look something like that in Figure 8. It is interesting to note that thematically, the LOEs and desired conditions of Figure 8: and the Ways and Ends of Figure 7 are comparable—perhaps because they strike at the root of the problems of the institutional aspects of space warfare or perhaps because the people developing them were educated and trained within similar institutions. Developers of strategy should always consider both possibilities.

---

*   Ibid., 7.

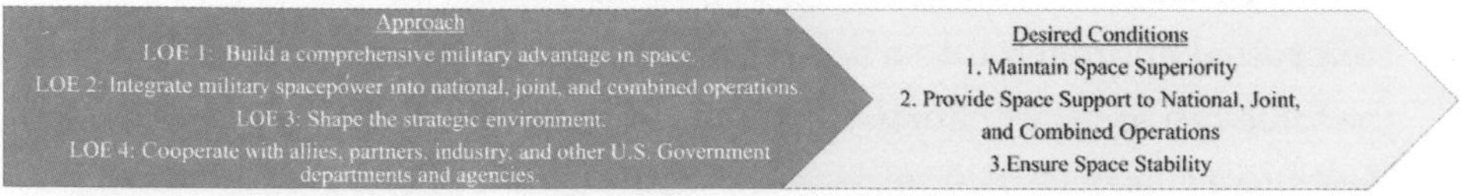

Figure 8: The Approach and Desired Conditions of the Defense
Space Strategy. *Source:* Created by authors.

The second point to note from the *Summary* is that multiple LOEs contribute to multiple desired conditions. In fact, each LOE seems to contribute
to each desired condition, an indicator of a well-integrated strategic concept.
It is similarly noteworthy that each LOE has objectives unique to that LOE.
The preceding objectives given for LOE 1, for example, are not objectives
within the other LOEs, indicating that perhaps the approach itself could
be subject to more rigorous integration. When LOEs share objectives—
when one objective contributes to multiple LOEs—there is opportunity for
simultaneous rather than parallel advancement of efforts. These efforts, of
course, require regular assessment, and because the strategic environment
will change, periodic revision. For any institution charged with creating the
future, accurately predicting what the world will look like a decade or more
from now presents a daunting challenge. This challenge is especially acute
for space operations because historical experience is limited. The final two
sections of this chapter more closely examine two manifestations of this
problem: the correlation of forces problem and the need for the development
of sound doctrine as a part of an institution's strategy.

Section 4: History and the Correlation of Forces Problem

Identified gaps in friendly force capabilities drive the military institution
to find solutions to meet the envisioned end states, while identified gaps
in knowledge of the enemy drive both short- and long-term intelligence
requirements. The shorter-term intelligence will affect the warfighting
aspects of the military more directly. Warfighting responses, by nature, are
more responsive to crises than the institutional elements. The long-term

intelligence requirements will have an effect on the more institutional aspects of the military, informing concept development, modeling and simulation, budgeting, acquisition, and the development of families of plans (groups of contingency plans to be executed by one or more combatant commands).

For the nascent space force, institutional processes—particularly acquisition processes—are of primary concern. Fortunately, there are available models for the acquisition process, but unfortunately, there is a dearth of models that can quantify the value of space to the joint fight—a fact that will inhibit acquisition efforts. As far as acquisition models go, the direct acquisition authorities of US Special Operations Command offer an intriguing model as a unique community without a dedicated "Special Forces Service" to perform its acquisitions functions. The Missile Defense Agency provides an alternative—or perhaps complementary—mode of an acquisition agency that develops capabilities with service inputs and then fields them to the services; this was the model for the Space Development Agency. A third model, the traditional service acquisition process, is still a distinct possibility as well, but the question of what is the best way to implement such processes remains an open one.

Whatever the combination of acquisition methods, institutions (generally, military services) must develop capabilities and implement them across the force to maintain warfighting effectiveness. They must decide what kind of capabilities and how much to invest in them. In the US system, capabilities fall into the DOTMLPFP framework (doctrine, organizations, training and education, materiel, leadership, personnel, facilities, and policy, or "Dot-Mil-P-F-P").

Generally, in the DOTMLPF-P model, capabilities are cheapest to implement starting with "D" and get more costly as the acronym progresses.* While materiel capabilities are often the most visible, to have a fully developed capability, the institution's capability development effort must address all aspects of DOTMLPF-P. In the implementation of a materiel solution, for example, there must be doctrine to tell the force how to employ the new

---

*   The authors would like to thank Lieutenant Colonel (Retired) Sam Fuller for sharing this observation.

equipment, new or modified organizational structures to man it, training plans to make the new organization effective, leaders and people who understand the fundamentals as well as what has changed, facilities to store and maintain the new equipment, and policy to allow for its employment in exercise and conflict. Some refining of the capability may be necessary when it reaches the warfighter—indeed, feedback and refinement are continuous processes of their own—but warfighters should not be forced to conduct capability development that the institution should have completed on their behalf. As the first element of capability development, the creation of doctrine signifies that the employment of the capability has been well thought out and is nested within existing doctrinal frameworks. While the next section deals specifically with doctrine creation, as a concept, doctrine causes some confusion among military practitioners and bears momentary discussion here.

Doctrine is distinct from strategy. First, doctrine is a tool proffered by the institution that has both institutional and warfighting uses. Although it often accounts for means and the way in which to employ those means (as in the notional fielding of a materiel solution discussed earlier), doctrine is not situation-specific or proscriptive but captures the how-to of common processes and best practices. Doctrine publications often focus on a given topic (operations or intelligence, for example), and these publications may combine ideas, but doctrine does not integrate these various ideas into a plan to be executed and assessed. With thorough and well-understood doctrine, the institution has primed the warfighting element to attain a common understanding. A strategist may then be confident in the performance of the force and integrate the various means and ways available into an executable plan. Like the other distinctions pointed out in this chapter, the distinction between strategy and doctrine is an important one for space forces moving forward, both for the purposes of practical implementation and for the development of a warfighting culture built around a common language. The implications of such a language will manifest themselves in future discussions of operational art, tactics, and their related subtopics.

A doctrinal language, of course, does not emerge ex nihilo. It depends upon the ideas that surround the culture creating it, and although anchored in historical study, it is also often anticipatory. Indeed, in the quest to anticipate what future capabilities are needed to overmatch an enemy, space warfare runs into a problem: the ability to correlate forces is not as complete as it would be for the air or maritime forces because of a lack of historical information and viable models. Since space warfare still exists primarily to ensure or to deny satellite support to military forces or civilian populations on earth, the question arises: what happens to terrestrial forces and civilian populations when their satellite capabilities are removed? To put it another way, what is the value of a satellite not in terms of dollars but in terms of warfighting capacity? How does space-generated information affect the battlefield, and how does its delay or loss affect the overall outcome of the battlefield? In considering this question, the long-held "satellites have no mothers" rationale breaks down.

We start from the position that space capabilities are mostly the providers of information and the transmitters of it. Space provides imagery, weather, navigational, signals intelligence, and missile launch information and transmits this information through communications satellites. The value of information provided by space systems cannot be overstated. In any military environment, information is essential for the conduct of operations—although the nature of the information employed may vary. Navigational information, for example, may be more important in an open desert without well-defined roads, but positioning information may be more important in a place like Korea where forces are maneuvering in a relatively small area full of vegetation where fratricide could be as great a danger as infiltration. How does the loss or delay of this information affect the overall outcome of terrestrial battlefields? Furthermore, since information affects human thought processes, it is difficult to measure the effects of lost or altered information on the battlefield. What is the effect of a successful misinformation or disinformation campaign? How might the battle have turned out differently if a piece of missing information had arrived on time? Questions like these make it difficult to balance antisatellite systems development and deployment. Is it better to remove an adversary's ability to image the battlefield or eliminate his ability to transmit this information to his own forces? Are there timing issues of when imagery should be denied or communication channels altered? What is then proper sequencing of ASAT employments to achieve such effects? How many do we need to achieve the desired overmatch?

The loss of a physical satellite is less consequential than the lost capability provided by that satellite. This lost capability potentially means the deaths of military personnel and even civilians—not due to the destruction of the satellite itself but due to the additional friction that arises from the loss of satellite communications, from the loss of missile warning, from the loss of precise positional knowledge. Just how gravely the lost capability could impact operations remains an open question. In ground combat, it is widely accepted that an armored force attacking an enemy armored force entrenched in a defensive position requires (as a planning factor and all other factors being comparable) an overmatch of 3:1. No such correlation factors exist for space warfare. How much does space-based intelligence, for example, whittle down the need for that 3:1 overmatch? How greatly does the loss of capability inhibit maneuver or increase expected casualties? A task force that can maneuver precisely, risking closer fires and enjoying greater situational awareness than it would in a GPS-degraded environment, may not need as much of an overmatch. Similarly, a space-air-ground task force with responsive satellite imagery may realize that the defensive position can be bypassed, altogether nullifying the need for an attrition-based correlation of forces model. The success of US forces in the Persian Gulf War and in the 2003 invasion of Iraq provide historical examples of space-enabled forces who enjoyed such advantages. Here we draw a distinction between space-enabled forces and space forces, and this is one of the most important characteristics of the interwar inflection point in which we find ourselves: space operations have surpassed their initial stage where they primarily enabled other-domain forces. Space operations have now become operations in their own right, capable of independent action. What does this mean?

Each of the combatant commands can employ their service-provided forces in whatever way best suits a given scenario. In some relatively simple cases, a naval patrol or an aerial show of force may attain the objective of the operation. Such actions mostly depend on single-domain forces. With the space force's expanding offensive capabilities for electronic and orbital warfare, space operations forces have some capacity to conduct the space equivalent of a patrol or a show of force. Importantly, new weapon technologies provide considerable advantages to those who use them first in conflict, fundamentally and unpredictably changing the correlation of forces at least for a period of time. In the siege of Constantinople, cannons demonstrated their superiority over walled fortifications, Civil War ironclads demonstrated superior survivability against wooden ships, and the maturation of airpower challenged the usefulness of the battleship. It is quite possible that the side that first employs offensive weapons against space systems in a conflict—whether on earth, on orbit, or within the electromagnetic spectrum—will "win" the space war, knocking off balance the powers who have heretofore employed their space forces in predictable ways.

Because military strategists should only rarely, if ever, allow forces to square off with an enemy toe-to-toe (and despite the acknowledged gap in historical and modeling data), it seems entirely reasonable that a small space force, even acting independently, could defeat a larger one. As in many historical conflicts—Scott in Mexico or the Israelis in the Yom Kippur War, for example—the possession of a larger force does not necessarily guarantee victory. Other factors like political will, strategy, morale, and culture all contribute to a force's effectiveness in combat. Given limited historical examples of a space force engaging in conflict, even the idea of victory or "winning" in space warfare is not clearly defined, and it is extremely difficult to assess how many or what kind of weapons systems the service will need to acquire. As an added twist, with the larger contributions of the joint force, it may even be possible to "lose" in space but still attain the desired policy objectives. Again, no correlation of forces models exist to accurately model this possibility. When they do, they will certainly be useful in informing both strategy and doctrine.

Section 5: Doctrine–
How the Institution Creates Itself

While institutional process is more complicated than a handful of boxes on a chart, the preceding schema emphasizes the importance of doctrine within the institution. As a fundamental element of the institution, doctrine is foundational to the education and training of the force (institutional functions) and to the employment of the force (warfighting). It should account for adversary doctrine as completely as possible to enable the force to understand the probable consequence of its decisions.

Through the lens of the education and training that they receive and informed by the situation around them, space personnel will develop the plans to implement capabilities across all levels of war. So, for warfighters to properly employ doctrine, they must have an intimate understanding of it. They must first have an appreciation for accepted practices for their own sake, but they must also be familiar enough with accepted practices to deliberately deviate from them when necessary. In a word, one may play jazz with doctrine, but one must be able to play the piano first.* Again, strategy is not doctrine, and that point bears special deliberation for the new space force because doctrine becomes a driver of service culture.

Based on the concept of strategy outlined in the previous section, doctrine is properly a means of the military institution to codify and purvey knowledge. It is a tool of military culture creation. As a product of the institution, it is necessarily somewhat static; bureaucratically, it takes time to reassess, refine, and reissue doctrinal publications. Apart from these practicalities, there is also the more philosophical question of how often a service (or the Joint Force or NATO) *should* revise its doctrine. It is true that, at some point, most doctrine becomes passé. After all, the French doctrine of 1914 was certainly different from their doctrine in 1919. Still, some doctrine seems to stand the test of time, with the caveat that no one can ever be quite certain when the test of time will fail. In fact, over time, institutions run the risk of following doctrine not because they believe it

---

* The authors attribute this turn of phrase to Lieutenant Colonel (Retired) Sam Fuller, Space and Missile Defense Command. Phone conversation with Jerry Drew, March 2019.

is inherently correct, but because institutionalized biases have led them to believe its correctness. This question of timing, biases, and correctness becomes a question not just for the writers and approvers of doctrine but for those who employ it.

For the space force moving forward and for other services that will continue to field space capabilities, the topic of doctrine requires the utmost criticality. When doctrine falls short or fails, the force will find creative nondoctrinal solutions to see them through the fight at hand, but these solutions are likely to be ad hoc, localized, and underinformed. The US Army found itself in this situation during the first years of the war in Iraq. Soldiers on the ground were implementing and sharing best practices based on experience, expert advice, historical examples, theoretical writings, and even older doctrine. The publication of Field Manual (FM) 3-24, *Counterinsurgency* in June of 2006 served as a milestone in the development and codification of thought toward counterinsurgency, but it was only one available tool. Of itself, FM 3-24 was not a strategy, nor was it intended to be one. Rather, it served more as a primer for practitioners at all levels of war to help them make sense of what was happening and to suggest best practices.

As an example of a doctrinal manual printed as a reactive publication, FM 3-24 symbolizes the ever-present tension between warfighting needs and institutional process. Ideally, warfighting needs should drive the institution, and in the case of FM 3-24, this happened. The arrival of the document more than three years after the Iraq invasion (almost five years after the Afghanistan invasion) provides an additional lesson. Despite the best efforts of the people within the institution, making sense of a problem and codifying it takes a significant amount of time. The continual revision and creation of doctrine provides the benefit of allowing the force to anticipate the next conflict. Ideally, one would have the doctrine on the shelf that was applicable to the next fight, which happened in the Persian Gulf War. In this case, rather than fighting the Soviets on the plains of Eastern Europe, as doctrine had envisioned, US forces fought a Soviet-style army in the deserts of Iraq. The doctrine of Air Land Battle was sufficient for this task.

Practically speaking, however, warfighting does not always drive the institution. In fact, it is quite possible for the opposite to happen. Consider, for example, a new piece of equipment that has been in development for a decade. Its initial conception and development may have stemmed from warfighting needs, but as in the case of the Army's Future Combat Systems, the assessment of these needs depended on a series of assumptions that proved faulty. To make matters worse for that program, the ongoing fights in Iraq and Afghanistan required the immediate attention of the institution, which had little use for a program whose aims were showing themselves to be more and more impractical. In the case of the Future Combat Systems, only a handful of subordinate programs made it to the warfighter. In the case of other long-lead capability development efforts, the technology is already well outdated by the time it reaches the force, yet the warfighter has no choice but to employ it anyway.

On a whiteboard at Fort Leavenworth's Command and General Staff College on any given day, one may see the following model for the creation of doctrine (Figure 9).*

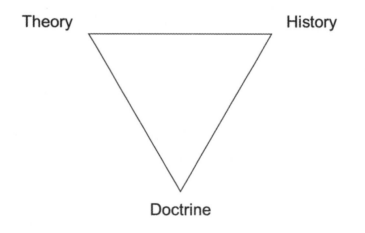

Figure 9: Base Model for the Creation of Doctrine. *Source:* Figure created by authors to depict a common model.

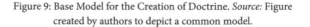

*   Although it is not possible to say where the inverted triangle model originated, it was taught to Jerry Drew by Dr. Dean Nowowiejski, the Ike Skelton Distinguished Chair for the Art of War at the U. S. Army Command and General Staff College in 2017.

This model illustrates that doctrine is the synthesis of theory and history. We may assume that this doctrine exists in an adversarial context that accounts for the theory, history, and doctrine of the enemy. While this process is true enough, the preceding model has three significant shortcomings. First, it does not depict the fluidity of the process. Second, it does not take into account the external factors that bear on the creation of doctrine. Third, it does not address the unique place of concepts as an intermediary step in the creative process.

While doctrine is often a product of theory and history, doctrine also influences history and theory, theory influences history and doctrine, and history influences theory and doctrine. The components are interactive. By way of example, Baron von Steuben's mixture of British drill regulation with elements of the Prussian system (doctrine) brought a professionalism into the Colonial Army that had a direct impact on the outcome of the American Revolution (history).* By the Napoleonic era, however, Prussian doctrine, despite modifications like adopting the French divisional system, was obsolete.† Napoleon's cataclysmic defeat of the Prusso-Saxon force at Jena-Auerstadt in 1806 (history) demonstrated this obsolescence, forcing the Prussian army to reconsider its entire system. Carl von Clausewitz's *On War* (theory) was an outgrowth of his experiences with the Prussian Army, but it was the ideas of his contemporary theorist Antoine-Henri Jomini that first found their way back into American doctrine, dominating the military thought of the Civil War period.‡

In addition to the interplay of theory, history, and doctrine, the preceding historical thread hints at some of the externalities that bear on the creation of doctrine, mainly individual, institutional, and cultural biases. These biases are not necessarily negative—in fact, they are probably unavoidable—but they will all influence the final product. The development of FM 3-24, *Counterinsurgency*, for example, manifested cultural biases that favored "winning the hearts and minds" through population-centric

---

*   Michael Bonura, *Under the Shadow of Napoleon: French Influence on the American Way of Warfare from Independence to the Eve of World War II* (New York: New York University Press), 42–43.

†   Michael V. Leggiere, ed., *Napoleon and the Operational Art of War* (Boston: Brill), 183.

‡   Bonura, *Under the Shadow of Napoleon*, 110–111.

counterinsurgency rather than through harsher historical methods like mass relocation or torture. Not only was population-centric counterinsurgency the most politically acceptable methodology, but it was a methodology that was not contrary to the ethical frameworks of the military services—a kind of positive institutional bias. The individual in charge of creating this publication was then-Lieutenant General David Petraeus, who brought his academic depth and practical experience to the problem set, inviting a host of experts to contribute—each with unique experiences, knowledge, and world views that would, in theory, be able to counter any negative biases through diversity of thought.

In the updated model (see Figure 10), cultural, institutional, and individual biases surround the creation of doctrine. No one type of bias corresponds to a single side of the triangle; all types of biases will affect theoreticians, historians, and doctrine writers alike, which is why awareness of these biases is crucial to creating a product that is as objective as possible.

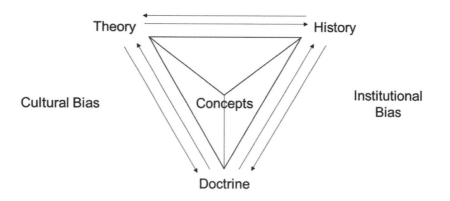

Figure 10: Modified Model of Doctrine Creation. *Source:* Adapted from the common model above by Jerry Drew.

Finally, the base model does not account for the very important role of warfighting concepts. Concepts emerge from an understanding of the current operating environment, an anticipation of the future operating environment, and an acknowledged need to modify current methods to meet future challenges. Concepts codify an approach by which the institution may continue to be relevant, and in this sense, they are part of something like an institutional theory/strategy within a service. They rely on long-term intelligence projections (the conceptual opposite of history), wargames, statistical analysis, and other means to try to validate the approach for its future usefulness. This task is fraught with uncertainty, but when concepts are validated, they may lead to doctrinal modifications—as well as other capability development efforts.

The preceding model of doctrine development does not depict—intentionally—the possibility of doctrine being created solely as an outgrowth of other doctrine. The army's current FM 3-14, *Space Operations*, provides something close to an example of this phenomenon, being largely a derivative of Joint Publication 3-14. This observation does not necessarily make the doctrine invalid from a conceptual standpoint. After all, service doctrine should be nested within joint doctrine—and to the extent practicable for US forces, synchronized with NATO doctrine—but the development of each publication should include a fresh look at the historical, theoretical, and doctrinal precedents underpinning that publication's development.

Conclusion

By the Department of Defense's definition, strategy deals with the implementation of policy at levels of the national government and includes all of the state's available instruments of power. In this sense, as discussed, the approved definition takes on the meaning of grand strategy. There is, however, a clear need to distinguish military strategy as an integral subset of grand strategy. The military strategic resides as an interface between grand strategy and the operational level of war and has two aspects, the warfighting and the institutional. The warfighting aspect depends upon the institution for its capabilities, and the institution bears the responsibility of fielding a combat-ready force.

In any professional conversation about the disciplines of war, agreed-upon definitions are necessary. Even with published definitions, however, meanings shift among different subcultures, and the lines are often blurry. The distinction between grand strategic and military strategic provides one example of a place where clear lines do not exist, but there are others. The levels of war, importantly, are not discrete ideas but suggest a delineation that, at some point, tips in favor of one concept or another. Similarly, the question of where policy ends and strategy begins remains an open one, one that is situationally dependent.

The schema for the military strategic (Figure 1) accounts conceptually for the interaction between these two aspects and their linkage to the operational level of warfare. Each aspect will necessarily develop its own strategies—an integration of ways and means to achieve desired ends.

As in the *Defense Space Strategy Summary*, ends may be referred to as "desired conditions," perhaps a more appropriate naming convention because the assessment of strategy and its progress are continual, rarely ever ending. Similarly, practitioners often refer to the set of ways as an approach. Whereas the available means limit the ways that the warfighting military strategist can employ a force, the institutional strategist faces a less rigid constraint. In fact, it is the duty of the institution to envision future approaches and develop the means that can turn those concepts into feasible alternatives for the warfighting commander.

Russian actions in Ukraine suggest the complexity of trying to unravel policy from strategy from tactical actions, and as the case of the US military's COVID response illustrates, the activities of one sphere are influenced by activities in the other spheres. Certainly, space as a warfighting domain has no shortage of policy issues and strategic considerations to explore, and each of these will have ramifications across the instruments of national power and the levels of warfare. These considerations will shape the future of capability development, including the creation of doctrine, which is fundamental to both service culture and warfighting effectiveness.

The policy and strategy issues offered here and a schema for considering the military strategic through the lens of space operations serve only as a starting point. This chapter is not a treatise on strategy or policy but a brief overview to set the stage for practitioners of war and warfare who will be implementing policy and strategy decisions, mostly at the operational and tactical levels. Ultimately, a lack of clarity at one level or a lack of understanding at another will degrade military effectiveness across the spectrum of warfare. As the means of creating shared understanding, the problem of language, then, continues across the operational and tactical levels, and even old ideas require explanation through the lens of space warfare. Sound doctrine will be a principal contributor to that endeavor, and despite the uncertainty of the future, earnest attempts to develop doctrine for future conflicts remain necessary.

Preparing for the future is a difficult task. Uncertainty over how a future war may play out is a problem, especially for space operations forces, who, with the creation of a unified command and a military service dedicated to space, are entering a new era of warfare. The correlation of forces problem and the need to develop sound doctrine are only two of the challenges facing both warfighters and the institution, but both are essential, and we cannot wait until we have more thorough historical examples to examine. The doctrine that is developed will form the basis of the language that is communicated to the force through education and training and that will allow practitioners to employ sound tactics to overcome the enemy—despite the inherent uncertainty. Subsequent chapters aim to develop and communicate additional language necessary to achieve these feats. To continue this discussion, we briefly leave the levels of war for a look at another framework—the principles of war, which do not fit neatly into any one level but are critical for the execution of warfare across all levels.

CHAPTER 2

Space Warfare and the Principles of War

Principles of War—Principles that guide warfighting at the strategic, operational, and tactical levels. They are the enduring bedrock of US military doctrine. See FM 100-5.
—Field Manual 101-5-1, *Operational Terms and Graphics*, 1997

Principles of Joint Operations—Time-tested general characteristics of successful operations that serve as guides for the conduct of future operations. (ADP 1-01)
—Army Doctrine Reference Publication 1-02, *Terms and Military Symbols*, 2016

Introduction

The version of principles of war adhered to within the US armed forces comes to us largely from the British Army officer Colonel J. F. C. Fuller, who was writing a century ago. From his studies of the campaigns of Napoleon, Fuller derived his initial list of principles in 1912, adding the final two in an article published in 1916: objective, offensive, mass, security, surprise, movement, economy of force, and cooperation.*

---

* J. F. C. Fuller, *The Foundations of the Science of War* (London: Hutchinson and Company, 1926), 13–14.

The contemporary American conception replaced movement with maneuver, removed cooperation in favor of unity of command, and added simplicity. Furthermore, the joint force has replaced the term *principles of war* itself, renaming the set the "principles of joint operations" and adding restraint, perseverance, and legitimacy as principles in light of the Iraq and Afghanistan experiences.* As the doctrinal language changes, so does our understanding of the concepts. Within the context of space warfare, an exploration of each of these concepts is necessary.

Mass

Mass (JP 1-02, NATO)—1. The concentration of combat power. 2. The military formation in which units are spaced at less than the normal distances and intervals. (Army)—To concentrate or bring together fires, as to mass fires of multiple weapons or units. (See also principles of war.) See FM 100-5.
—Field Manual 101-5-1, *Operational Terms and Graphics*, 1997

It is tempting to believe that some of the principles of war are not applicable for space warfare. "Massing" seems like a particularly tellurian principle, depending as it does on the physical placement of overwhelming force at the decisive point of a military operation.† For space operations, a belligerent may mass antisatellite missiles and orbital assets in such an intuitive way, but the opportunity also exists to think about massing in less conventional ways. Space operations forces, for example, may mass offensive electronic warfare effects or space domain awareness capabilities, or conceptually, they may mass space support for an operation, giving a supported commander

---

* In the most recent definitional publication—*Department of Defense Dictionary of Military and Associated Terms* (2018) and Army Doctrine Reference Publication 1-02, *Terms and Military Symbols* (2016)—most of the principles themselves are not defined terms. The exceptions are that *objective*, *unity of command*, and *security* are defined in both publications; *economy of force* is defined only in the DOD dictionary; *mass* is defined in the army publication only in reference to massed fires; and the Army defines simplicity only with reference to sustainment. As a result, we have defaulted to older doctrinal publications for useful definitions.

† Indeed, massing at the decisive point was Jomini's fundamental principle of war. Unlike Fuller, who only studied Napoleon a century later, Jomini observed the emperor's armies firsthand.

overwhelming communications, navigation, or missile warning support at a specified moment in the battle. It is even possible to discuss massing in terms of massing a joint function—massing command and control, for example. Whatever the combination, space forces will stand the best chance of mission success if they are able to mass multiple types of weapons systems on orbit, in the electromagnetic spectrum, and on the earth's surface in a way that is synchronized with joint operations in other domains.

Massing orbital warfare assets could potentially be done in two ways: massing against a single target or massing within a "choke point" (choke points contain multiple satellites that share orbital parameters but are not necessarily physically near one another; more on this concept and how to map them in Chapter 4). While the prospect of massing a host of coorbital antisatellite weapons against a single target certainly sounds intriguing, such an attack may be inadequate to achieve a significant and lasting military advantage unless the target is of a uniquely high value. For satellites that work as part of a constellation, destruction of one satellite does not necessarily equate to destruction of a total capability.

Consider the GPS constellation. Eliminating only one or two GPS satellites might not significantly reduce the overall navigational accuracy on the battlefield. To make GPS truly ineffective, an adversary would have to "mass" its attacks to eliminate a significant portion of the GPS constellation at the appropriate orbital locations and at a precisely determined moment in the conflict—that is, within one or more choke points. Limitations of sensor coverage may pose a challenge to determining the decisive location and time to execute the attack, but the orbital motions of most satellites are fairly predictable, which would reduce the risk of the attempt. Recovering from such an attack would be extremely difficult, but given the vastness of space, the limitations of sensor networks, and the newness of space warfare, simply assessing the effectiveness of the attack may be difficult, and assessing the compounding effect of multiple engagements may be next to impossible. As a result, the enemy may have a difficult time exploiting his success, buying time for recovery operations. In a prolonged conflict, piecemeal destruction may provide the time necessary for the opportunity to reconstitute lost capabilities.

Within the range of electromagnetic spectrum activities, several opportunities for massing exist. In support of a joint operation, for example, a commander of space forces may decide to mass satellite communications support during a critical phase of the campaign in support of a field army or a joint task force. While not strictly satellite communications, GPS operators accomplish a similar feat by adjusting orbits to "mass" their positioning accuracy for particular theaters by modifying satellite orbits or increasing the strength of the provided signal. We might think of these acts, especially when combined, as massing the function of command and control within the space domain in order to allow the joint force to mass the functions of maneuver and fires in the other domains. In a similar way, space forces may mass sensor coverage (including ground-based and space-based sensors) against a threat either for intelligence gathering or for protection purposes. Being able to gather critical situational awareness of a coming attack—whether against friendly forces or against the space forces of another country—will inform not only the military but also the diplomatic, information, and economic response options of the nation. Such massing activities as these depart from the Napoleonic concept of massing, but then again, Napoleon did not have to consider what it means to mass in the electromagnetic spectrum.

Napoleon did, however, have to worry about prepositioning and forward stationing his forces so that they could mass at the appointed time. For an expeditionary military like that of the United States, the ability to mass space capabilities is highly dependent on both prepositioning and forward stationing (i.e., basing). Three factors bear consideration in these basing decisions. First, many of the existing equipment sets for space operations are immobile (GPS satellite control stations, for example), fixed at locations that may make them irrelevant to certain operational scenarios and highly vulnerable in others. Second, as with tanks and armored personnel carriers, a person trained to operate one type of space operations equipment may be completely untrained to operate a conceptually similar piece of equipment, thus limiting the ability of the force to cross-level personnel during a conflict. Third, even if equipment becomes mobile and personnel become cross-trained on multiple systems, the anticipated demand on transportation assets in a high-intensity conflict will severely inhibit moving to the most advantageous location in time to affect the outcome of the conflict, especially if space systems are misunderstood and considered a lower priority to the expected conflict.

Whether on orbit, within the electromagnetic spectrum, or on the earth's surface, there are limits to how effectively space operations forces may be able to mass during a conflict—even if they want to. Many space weapon systems could gain advantage by adhering to the principle of low visibility, and in many circumstances, it will benefit a belligerent to attempt to achieve low probability-of-verification and low probability-of-attribution of attacks. Assuming, however, the political will is available to mass forces, there are only limited numbers of personnel and equipment sets available for each kind of space operations system. There may not be sufficient forces available to achieve continuous space control, or forces may not be in the proper geographic region—either on land or in space—to act at the moment of crisis. Available forces may be able to mass, but their mass may not be large enough to achieve sustained or overwhelming space control, and the enemy may be able to reconfigure their communications, for example, to avoid serious detriment. In addition, it is difficult to maneuver satellites to new orbital locations to concentrate their effects. This takes time and depletes precious orbital fuel, limiting the life of these satellites.

If the weight of the punch is the operative idea behind mass as a principle, preparing for the punch is a second, the timing of the weighted punch is a third, and the way in which the punch is thrown is a fourth. The idea of combined arms applies here just as it would for a ground force. An armored corps, for example, could mass solely with its tanks. It is likely to be more effective, however, if it simultaneously masses its artillery, its air support, and its logistical support. Space forces could mass only on orbit or only in the electromagnetic spectrum independent of military operations in other domains, but they will be most effective if they are prepared to throw a punch from the most advantageous location, at the most advantageous time, with the most creative tactics, operational art, and strategy.

Economy of Force

Economy of Force—The judicious employment and distribution of forces so as to expend the minimum essential combat power on secondary efforts to allocate the maximum possible combat power on primary efforts. (JP 3-0).
—*DOD Dictionary of Military and Associated Terms*, 2018

Economy of Force—The allocation of minimum-essential combat capability or strength to secondary efforts so that forces may be concentrated in the area where a decision is sought. Economy of force is a principle of war and a condition of tactical operations. It is not used to describe a mission. (See also main effort.) See FMs 7-30, 17-95, 71-100, 71-123, 100-5, and 10015.
—Field Manual 101-5-1, Operational Terms and Graphics, 1997

The preceding doctrinal definitions of *economy of force* address the tension between the principle of mass and the principle of economy of force. In this conception, the main effort receives the benefit of mass while supporting efforts receive only what they need to accomplish their missions (which may be mere diversions). Thus, in the doctrinal definition of the concept, the entire principle revolves along the employment of physical force. In the original concept of economy of force, however, J. F. C. Fuller considered not only the economy of physical force but also the economy of mental and moral forces. As with the principle of mass, economy of force bears additional explanation within the context of space warfare.

As discussed in the section on mass, the prevailing characteristics of the space warfare operating environment—from a US perspective—include a limited pool of available forces, the responsibility for numerous missions, the conduct of operations all around the world, and the political and diplomatic consequence of deploying military forces. Under the current method of employment, US space operations forces operate in an economy of force role; the minimum essential combat capability is allocated. By the numbers, Air Force Space Command, at the dawn of the space force, consisted of more than 26,000 personnel (of an active air force of about 328,000).* US Army Space and Missile Defense Command (USASMDC) included approximately 2,800 personnel.† In both organizations, these numbers include research and development, capability development, institutional training, and education personnel, so the actual number of people conducting real-world operations was far less than the total—in USASMDC's case, about 800 of the total.‡ Although the preceding doctrinal definition from 1997 states that "economy of force" is not used to describe a mission, in practice, such is often the case. What's more, space operations have heretofore been economy of force missions for the joint force—a position that would not, from the standpoint of traditional military thought, make space operations deserving of their own force. After all, independent services exist—at least partly—because they can fulfill national policy objectives with some level of independence.§

The label of "economy of force mission" is not a negative one. Indeed, space operations forces operate multiple global missions with a relatively small amount of force structure. Missions such as missile warning and satellite communications, without question, must continue, albeit with a sea change in professional ethos. In the relatively benign space environment of the post-Cold War era, the notion that space professionals were service providers was largely sufficient. With the designation of space as a warfighting domain, however, the service-provider mentality must depart, taking with it the conception that space operations forces can only fulfill the economy

---

* US Air Force Space Command, "Air Force Space Command," July 19, 2018, accessed September 20, 2020, https://www.afspc.af.mil/About-Us/Fact-Sheets/Display/Article/249014/airforce-space-command/; US Air Force, "Military Demographics," January 1, 2020, accessed September 20, 2020, https://www.afpc.af.mil/Portals/70/documents/03_ABOUT/Military%20Demographics%20Jan%202020.pdf?ver=2020-01-27-093137-550.

† US Army Space and Missile Defense Command (USASMDC), "2020 Global Defender: A Guide to USASMDC," Undated, accessed September 20, 2020, https://www.smdc.army.mil/Portals/38/Documents/Publications/Publications/SMDC_2020GlobalDefender_031620_v508_Final.pdf. By comparison, the Third Infantry Division is about 14,000 soldiers.

‡ Ibid.

§ One may also argue that independent services exist to allow for easier management of the bureaucracy.

of force role. Being able to mass, however, requires investment, not only in equipment, but also in organization, training, and culture.

While massing is possible in nontraditional ways, as discussed in the preceding section, the space force is currently investing in a series of satellite jammers to fulfill the goal of being able to mass electromagnetic effects.* As these systems come online, planning for their deployments and employments will need to be integrated within an overall framework—either one that uses them in concert (mass) or one that piecemeals them out to other commanders (economy of force). Such low-density assets should be employed purposefully at all times, from reinforcing positions, and with the requisite intelligence, protection, and sustainment forces behind them—functions that themselves are economized based on priority. These space operations requirements imply the need for robust targeting preparation, the incorporation of weapons with different phenomenologies, and an idea of how to optimally employ such systems—even in delaying or limited deception activities—in the face of competing and dynamically evolving priorities. Even the "service-provider" missions like SATCOM require long-term investment in such areas, particularly with regard to cyber protection. Figure 11 illustrates the extent of how cyber-attacks could influence space operations whether by attacking the industrial base, ground control facilities, or on-orbit assets. In effect, any part of the space operations architecture that is dependent on computers or communications links is vulnerable.

---

* US Space Force, "21st Space Wing Squadron Poised to Receive First Space Force Weapon System," January 31, 2020, accessed September 20, 2020, https://www.spaceforce.mil/News/Article/2075953/21st-space-wing-squadron-poised-toreceive-first-space-force-weapon-system.

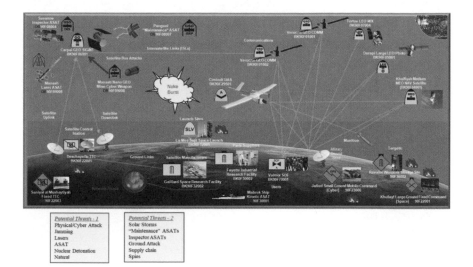

Figure 11: Notional Space Warfare Attacks. *Source:* Adapted from Paul Szymanski, "How to Win the Next Space War: An Assessment," *Wild Blue Yonder Online Journal* (March 2020). https://www.airuniversity.af.edu/Wild-Blue-Yonder/Article-Display/Article/2981831/how-to-win-the-next-space-war-an-assessment/.

Within the entirety of the joint force, space operations are likely to be subjected to economy of force constraints for the foreseeable future. However, whether within SPACECOM or as part of another combatant command's efforts, the functions of intelligence and protection for space operations should only be economized on rare and very deliberate occasions. Space operations forces, like forces in other domains, are anxious to deploy and execute their missions, but missions—especially those that rely on very limited personnel and materiel pools—are less likely to be successful when the concept of how to support them is underdeveloped.

Clarity of thinking was the cornerstone behind Fuller's ideas on the economy of mental force. "If our thoughts are chaotic," he wrote, "so also will our actions be chaotic."* In the employment of space operations forces, then, we should all strive to come to abundantly clear agreement on how the forces are to be employed, who is responsible for their operations (or parts of their operations), and what forces or functions are being economized and why.

---

*   Fuller, *Foundations of the Science of War*, 204.

Clear doctrine, well-written orders, and critical thought are just a few of the tools available to achieve economy of mental force. As Fuller says, "numerical superiority is only a special interpretation of the meaning of strength."* It may be sufficient to win the battle—but economizing in the moral, mental, and physical spheres will lead to a surer victory.

Security

> **Security**—1. Measures taken by a military unit, activity, or installation to protect itself against all acts designed to, or which may, impair its effectiveness. (JP 3-10)
>
> —DOD *Dictionary of Military and Associated Terms*, 2018

The goal of security measures is to protect the force so that it can preserve combat power to achieve its objective. The problem of security is inherently a multidomain one. Within the context of land warfare, an army division must protect itself from enemy infantry, armor, and artillery; from air and missile attack; from chemical weapons; and from saboteurs and spies and manage its electromagnetic signature. It must do all of these things with limited resources. The implications are that some units will not receive engineer support to help them improve defenses, some will not receive augmentees to help man their perimeters, and some will not fall under the air defense umbrella. In relation to the constraints of economy of force, security, therefore, is very much a question of what *not* to protect.

For space warfare, the problem is a similar one. Ground segments—whether for command and control, launch, or industry—require protection from a myriad of threats. Even sites in the United States or in friendly nations may be subject to insider threats, to external surveillance, or to cyber-attack. For this reason, operational security is an essential part of the overall security picture. In a battlefield environment, a forward space element would face all the same threats as an army division yet, without organic protection assets, remain dependent upon the unit to which it is

---

*   Ibid.

attached. The link segment, too, requires protection. Encryption, frequency hopping, monitoring, defensive cyber operations, and diversifying the communications plan are a few ways to ensure the survivability of data lines of communication. Keeping the ground and link segments secure inherently keeps the space segment secure, and until on-board artificial intelligence becomes standard on satellites, this will remain the case. In the past, critical satellite systems incorporated some survivability/resiliency features, such as nuclear radiation protection, command interference detectors, or sensor shutters to protect against laser attack. If not so specialized, most satellites have some inherent and rudimentary protection measures such as maneuvering away from approaching satellites, pointing their high-power antennas toward threats, or simply spinning or randomly changing attitude to counter grappler attacks. Finally, fundamental to the survivability of orbital assets is a network of surveillance sensors—both on the ground and in space—with the necessary algorithms and intelligence infrastructure to assess when satellites are in danger. In the future, satellites may include on-board sensors to provide warning of attack or self-defense capabilities to respond to one, but this practice is not currently the norm.

For the joint force, the security of space operations has significant implications. Given the dependencies of the joint force on satellite communications, for example, a security failure in the space enterprise could quickly ripple across the globe, affecting the fighting effectiveness of every combatant command. Active protection of space assets, then, is a paramount concern and may include the allocation of resources to keep safe critical ground control nodes, space choke points, launch locations, and portions of the electromagnetic spectrum. Another way to mitigate the security vulnerabilities induced by space dependencies is to develop alternative capabilities that perform the same function. This approach includes having secondary and tertiary means of obtaining and processing space domain awareness data, maintaining critical communications, and continuously assessing enemy strategy, tactics, doctrine, organization, and intentions. Alternatively, the joint force may choose to reduce its dependency on space-based capabilities within practical limits. The US Navy, for example, still trains with the sextant to reduce its dependency on GPS, the US Army with map and compass for the same reason. Ground-based communications, including fiber, provide an alternative to satellite communications, and manned aircraft, drones, and airships can provide localized battlefield surveillance. Radars cannot entirely make up for satellite-based missile warning, but with the appropriate allocation of resources, they may very well protect the most crucial components of the joint force.

Surprise*

Surprise—One of the nine principles of war. The enemy is attacked at a time or place, or in a manner for which he is unprepared and which he did not expect. See FM 100-5.
—Field Manual 101-5-1, Operational Terms and Graphics, 1997

Like the pair of mass and economy of force, security exists as a relation to the principle of surprise, and in this case, the principle follows the meaning of the word's common usage. For a professional service member, surprise is among the worst of things. Not only is it potentially catastrophic to the mission, but it also lays bare a failure to anticipate—a failure which professional soldiers get paid to avoid. Surprise means we have been outwitted, and that is not acceptable.

If security is the more defensive facet (Clausewitz might say the "negative" aspect), then surprise is the more offensive or positive facet. Like the other principles, surprise can exist at all levels of war. The attack on Pearl Harbor (as part of the larger Japanese offensive), with its diplomatic deception, its cross-Pacific movement, and its calculated political effects, provides an example of strategic surprise. The Battle of the Bulge, with its military deception, its unanticipated avenue of approach through the

---

*   Value of Surprise: A National Defense University study looked at the major conflicts of the twentieth century up to the 1960s. They categorized battles in these diverse conflicts according to whether the defenders were surprised by the attacking force. After assessing casualty ratios, they found that being able to surprise your adversary improves your casualty exchange ratios by 8.5 times. Certainly, surprise in any conflict is a key force multiplier. This is one of the main impacts of space systems: the eyes and ears in space minimize the ability of forces to conduct surprise attacks.

Ardennes, and its unlikeliness (based on available Allied intelligence on German strength) provides an example of operational-level surprise. In the history of warfare, tactical surprises are the most numerous and involve, as one would expect, taking advantage of the unexpected—an attack from an unusual angle, at an unusual time, with an unusual force. In the cases of Pearl Harbor and the Bulge, both the Japanese and the Germans initiated offensive action but went on to lose the war. In a similar vein, it is worth considering if the side that first employs an offensive in the space domain will "win" in the long run or if they will simply gain temporary advantage. Due to the impossibility of armoring satellites against hypervelocity antisatellite weapons and the difficulty of detecting and avoiding incoming ASATs, it might very well be that the side that attacks first will win in space. As with Cold War nuclear weapons, however, the temptation to strike first and the inability to defend against such a strike may also make space warfare unstable for escalation control.

Indeed, the realities of the hypertransparent operating environment with global satellite coverage, near real-time data relay, and mobile devices in every pocket may lull us into the false surety that surprise is impossible. Given the previously discussed difficulties of knowing what is happening in space, however, limited space domain awareness assets, and limited capacity to monitor the electromagnetic spectrum, surprise remains a distinct possibility for both friendly and enemy forces. Compounding the uncertainty is the possibility of unknown capabilities, unknown weapons that adversaries may have developed in secret and which may be difficult to detect or attribute, if employed. Even if their existence is known, the way in which an adversary ultimately employs them in a conflict (that is, their tactics) may come as a surprise. In the information environment, even the threat of a weapon (existent or not) can cause surprise. It is important to remember that adversaries will have the same questions about US forces. Are the existence, location, or employment techniques of US weapons known to an adversary, or does the potential for surprise still exist? Furthermore, is the use of these weapons detectable or attributable? Even if their existence is known, timing, tempo, or the threat of a weapon use can still enable surprise.

Unity of Command

> **Unity of Command**—The operation of all forces under a single responsible
> commander who has the requisite authority to direct and employ those forces
> in pursuit of a common purpose. (JP 3-0)
> —DOD *Dictionary of Military and Associated Terms*, 2018

The reestablishment of a combatant command for space in 2019 (the original
US Space Command existed from 1985–2002) brought significant questions
about the command and control of global space operations forces. In some
cases, theater commanders had been granted command authorities over
forward forces with some level of involvement in their operations. This
splitting of responsibilities led, on more than one occasion, to a lack of
clarity on just which commander was responsible for which aspects of the
mission. An in-theater commander might for example, dictate priorities
for the space mission itself (a partial tactical control), a garrison or base
commander may have force protection or logistics responsibilities (partial
administrative control), another commander administrative responsibility
(administrative control), and another commander the authority to displace
or relocate the force (operational control). In these cases, the commander
of the actual mission unit—perhaps in rank as low as a captain—may have
been required, depending on the circumstance, to answer to a variety of
majors, lieutenant colonels, colonels, and generals. The result was nearly
always a lack of common understanding among the multiple authorities—a
particularly dangerous scenario in times of crisis.

The new Space Command promised an opportunity to rectify some of
these incongruities, but the process is lengthy and ongoing. What is more,
since there has long been confusion over command relationships and
administrative responsibilities, it seems that each person involved in the
discussion has a different vision of the situation's reality—to say nothing
of the different visions of what the command relationships should be. Two
institutional realities lie behind the difficulty of command relationships,
which the space force must conquer if it is to be effective. First, joint doctrinal definitions of command relationships are vague, leaving out specifics
on the authorities conferred by specific relationships. Army doctrine on

command relationships goes further, explaining the specific duties and responsibilities linked to each type of relationship. Joint and army uses of some common terms differ—for the army, for example, a support relationship is not a command relationship, but it is in joint doctrine—and space operations personnel must understand these distinctions because they will be the ones responsible for communicating the plans to employ their forces. The second institutional difficulty is in the training aspect; command relationships are difficult to teach, largely, it seems, because relatively few professionals have the opportunity to wrestle with their nuances. With these questions, however, space professionals must be extremely diligent because, as will be discussed in the section on command-and-control graphics and in the section on rules of engagement, space missions are often global, often capable of supporting multiple combatant commands (even at the same time), and may depend on the authorities of the president or the secretary of defense.

Since the commander of US Space Command holds combatant command authority (COCOM) over most space operations forces, achieving unity of command for space operations—or some degree of it—rests largely with this individual. In tension with this desire is the reality that other theater commanders like to have control over the resources within their theaters. Mature combat theaters like US Central Command may have the resources, expertise, and (more or less) clear delineation of responsibilities to command and control some space operations. The benefit of this approach is that space operations can more readily integrate with other theater operations, including information operations and intelligence collection—possibly even supporting Space Command objectives in the process. The downside to this approach, however, is that a theater may monopolize an asset, preventing its employment for the benefit of other theaters. If the theater commander fails to appreciate the importance of space to the conduct of their operations, the sin is all the greater, because the asset may sit underutilized.

Below the echelon of combatant command, the issue of unity of command rears its head again—this time in the context of specific space missions. Take, for example, the SATCOM mission. Historically, army, navy, and air force personnel contributed to this mission under a responsible commander. The air force "flew" the communications satellites and managed some bandwidth (mostly in the extremely high frequency range), the army managed the majority of the military's satellite communications bandwidth (about 80 percent, mostly in the super high frequency range), and the navy specialized in ultra-high frequency. While these different components nominally answered to the commander of a joint warfighting command for space, as a practical matter, this individual bore responsibility not just for SATCOM but for all space missions—a tall order requiring depth of expertise in each mission.* As a result, it appeared that unity of command was achieved—at least organizationally, but breadth of responsibility for missions without like dependencies created a control nightmare. Subordinate commands for the various missions would alleviate this problem but would require additional overhead in the form of expert commanders and staffs for each mission—force structure that the space force may be able to provide.

At echelon, then, the need for integration not only of all space operations but of *all* operations has never been greater. As an integral part of all-domain operations, commands must strive to integrate space operations with air and ground planning, with information warfare activities, and with intelligence collection, just to name a few. Target development and allocation must give sufficient consideration to space targets in a way that departs from the "classic" targeting process, and these, of course, must be traceable to friendly objectives as they relate to the enemy situation. Above all, commanders must understand the importance of space operations to their fight and understand what authorities and responsibilities are granted to them.

---

*   Originally called the Joint Force Component Command-Space (JFCC-Space), this organization took up the mantle of space warfighting after the dissolution of the original US Space Command. No longer deemed worthy of a unified command of its own, space operations, from 2002 to 2018, fell under US Strategic Command, the organization primarily responsible for the employment of US nuclear forces. As one may imagine, the nuclear mission always took priority, which became an argument for the reestablishment of a Space Command. Later, when still under US Strategic Command, JFCC-Space was renamed the Joint Force Space Component Command (JFSCC, voiced as "jiff-sick"), and still later, with the recreation of Space Command, it was renamed the Combined Force Space Component Command ("siff-sick").

Maneuver

> **Maneuver**—1. A movement to place ships, aircraft, or land forces in a position of advantage over the enemy. 2. A tactical exercise carried out at sea, in the air, on the ground, or on a map in imitation of war. 3. The operation of a ship, aircraft, or vehicle to cause it to perform desired movements. 4. Employment of forces in the operational area, through movement in combination with fires and information, to achieve a position of advantage in respect to the enemy. See also **mission; operation.** (JP 3-0)
> —*DOD Dictionary of Military and Associated Terms*, 2018

As the preceding litany of definitions attests, maneuver takes on different meanings for different audiences. As a principle, and in keeping with the Napoleonic interpretation, maneuver is not simply moving forces from one location to another. More precisely, it is moving forces in relation to an enemy to gain an advantage and employing enablers to improve the chances of success.

As in nearly all military space operations, one must consider maneuver in all segments of the space domain (satellites, terrestrial receivers, antisatellite systems, whether space-based or terrestrial-based) and in the electromagnetic spectrum, including directed energy beams.

First, and perhaps most intuitively to the common military experience, space supports the maneuver of the other domains. A fleet of ships, for example, relies on satellites for communications, missile warning, intelligence, weather, and navigation (for movement of the ships but also for employment of munitions like cruise missiles). As such, a fleet is a self-contained fire and maneuver entity that can be further augmented by space capabilities, like ground-based antisatellite missiles or lasers, satellite jammers (in the sense of offensive space control), or communications monitoring (in the sense of defensive space control). In the fleet scenario, the antisatellite missiles and the satellite jammer serve as a fires element, while the communications monitoring provides, in the electromagnetic spectrum, a function something like that of a counterbattery radar. In this way, space fires can be integrated into surface maneuver.

The second—and perhaps most obvious—aspect of space maneuver is the orbital aspect. Satellites are constantly in motion, and a satellite's ability to frequently conduct large, small, or continuous maneuvers may enable positions of advantage, keep adversaries guessing about friendly intentions, and complicate his targeting solutions. Furthermore, frequent movements, movements outside of known sensor range, or entry into unusual orbits compounds this difficulty.

It may not matter how plentiful or how brilliant an adversary's weapons systems are if they cannot find or reach your critical space systems. While such movements are prudent and essential, their movements can only be considered maneuver in the warfighting sense if they are in relation to the enemy and coupled with fires. Satellites of the general military support variety are not designed for excessive repositioning, on-board fuel reserves remaining, along with battery power, the most significant constraints in spacecraft engineering.

Some satellites, however, are designed to maneuver in relation to other spacecraft. Historical examples include Project SAINT and the Geosynchronous Space Situational Awareness Program.* Another such program, the Air Force Research Laboratory's ANGELS program, employed small satellites to provide "a clearer picture of the environment around our vital space assets."† The potential spacecraft capabilities suggested by the Defense Intelligence Agency in their 2019 report *Challenges to Security in Space* open up the possibility of an entirely new conception of fire and maneuver in

---

*   Jennifer Leman, "Why Is This Russian Satellite Stalking a US Spy Satellite in Orbit," *Popular Mechanics*, February 5, 2020, accessed September 20, 2020, https://www.popularmechanics.com/space/satellites/a30767105/russian-satellite-stalking-us-spysatellite/; Colin Clark, "New Spy Satellites Revealed by Air Force; Will Watch Other Sats," *Breaking Defense*, February 21, 2014, accessed September 20, 2020, https://breakingdefense.com/2014/02/new-spy-satellites-revealed-by-air-force-will-watch-othersats/; "Geosynchronous Space Situational Awareness Program," Air Force Space Command, March 22, 2017, accessed September 20, 2020, https://www.afspc.af.mil/About-Us/FactSheets/Article/730802/geosynchronous-space-situational-awareness-program-gss-ap/#:~:text=Geosynchronous%20Space%20Situational%20Awareness%20Program.%20GSSAP%20satellites%20are,as%20a%20dedicated%20Space%20Surveillance%20Network%20%28SSN%29%20sensor.

†   ANGELS stands for Automated Navigation and Guidance Experiment for Local Space. "Fact Sheet: Automated Navigation and Guidance Experiment for Local Space (ANGELS)," Air Force Research Laboratory, July 2014, accessed September 20, 2020, https://www.kirtland.af.mil/Portals/52/documents/AFD-131204-039.pdf?ver=2016-06-28105617-297.

space. An adversary could, for example, employ an orbital laser weapon to blind a target's optical sensor (fires) while a second satellite with a robotic mechanism maneuvers toward the same target. Repeating similar tactics throughout orbital space in a synchronized manner hints at the possibility of an operational-level offensive in space. What is more, one might combine such an operational-level offensive in space with the activities of the fleet to enhance the chances of the main effort's success.

There is a caveat to anyone longing for such offensive action on orbit. Since it is very difficult to maneuver a satellite at the last minute (especially changing its orbital inclination), immediate space combat is only possible with nearby resources. Given current technological limits, transconflict redistribution of space forces to reinforce those under attack is infeasible. Thus, preconflict positioning of space assets is possibly the most important aspect of space military strategy. The goal may be to command the "high ground," to dominate the "high ground" from the earth, or to employ some combination of the two to achieve a measure of space superiority. To enable on-orbit conflict, it will also be necessary to take advantage of critical orbits, launch corridors, and communications paths, possibly even denying those to an adversary to reduce their freedom of action. It will be necessary to have primary and alternate positions for each of these capabilities, shifting activities to avoid predictability and to enhance survivability. Historically, the United States has been engaged in such defensive positioning since Explorer I through the introduction of ever more advanced space systems to serve as eyes, ears, and warning bells in the sky.

The third—and least intuitive—concept for space operations maneuver exists within the electromagnetic spectrum itself. SATCOM transmissions, for example, are physical waves that move through air and space. The movement of the wave itself (polarization, modulation, beam shape) can be manipulated for better survivability, and the same ideas also apply to laser communications capabilities (ground-to-space and space-to-space). The transmission itself can also move, from ground station to ground station, from satellite to satellite, from transponder to transponder. One signal could feasibly hop to a new pathway (or multiple new pathways) if under observation or attack. What's more, the data itself could be transmitted via many different means, originating, say, on a UHF satellite but transferring to an SHF satellite via a fiber connection or point-to-point radio transmission. Imagine now that these transmissions are synchronized with electronic warfare or cyber activity—the "fires."

Imagine that an enemy begins interfering with a critical satellite signal path. Aware of the threat, friendly space operators dynamically shift the transmission to a new transponder on the same satellite, but the enemy is able to find the signal again. The SATCOM team reaches out to the cyber and electronic warfare teams to enact a premade fire and maneuver concept. The SATCOM team begins the relocation of its critical signal while the cyber and electronic warfare teams begin attacking the portion of the enemy's command and control infrastructure that is controlling its jammers. In turn, the enemy begins a new set of cyber-attacks to try to negate the friendly force's capabilities. Now imagine similar scenarios happening dynamically across a subset of thousands of the most important SATCOM signals. This is another tall order but one that—despite its difficulties in relatively peaceful times—would become all the more challenging during combat.

A truly powerful concept for space operations maneuver would combine all elements—integrated planning for all-domain maneuver, an orbital maneuver plan, and an electromagnetic maneuver one. These plans are all related. Functionally, however, they all involve different but mutually supporting sets of expertise.

In the first element, an intimate understanding of the particular form of domain warfare (air, land, maritime, cyber) and its relation to a particular aspect of space operations (SATCOM, PNT, MW, ISR, space control) is most necessary. Historically, the army has charged its Army Space Support Teams with this role, and the air force has placed space weapons officers inside its Air Operations Centers to perform a similar function. Following the army model, the Marine Corps is developing Marine Space Support Teams, but at present, these are specially trained organic staff, rather than expert teams brought in from an external organization. The navy provides exceptional additional space training and education—through the Naval Postgraduate School, for example—to a handful of line officers, but there is no space-specific career field within the service, and officers trained in space are likely to serve in only one or two space-specific assignments. Nonetheless, that service's dependence on SATCOM and their historical relationship with signals intelligence ensures that they have some of the top experts in those missions.

The second element, the orbital maneuver element, may be what most people think of when they hear about military space operations. Based off of the DIA's report, this seems a likely area for mission growth in the near future and may be particularly active during competition phases of a conflict. However, it is worth asking at what point such systems would be able to significantly affect a large-scale conflict. Indeed, capabilities such as these would likely be the first targets of ground-based jammers, lasers, antisatellite missiles, or enemy satellite weapons. Given the sensitivities of such systems, it is unlikely that a belligerent would even acknowledge the damage, as their orbital capabilities remain closely guarded, even from allies. Furthermore, based on the constraints of launch and orbital dynamics timelines, optimally employing them would require significant lead times, and the assumed critical orbits, launch corridors, and communications paths envisioned prior to conflict may differ drastically from those that emerge from an unanticipated combat situation.

Even without optimal positioning, however, it is still possible for orbital attacks to occur within a relatively short period. A series of simulations conducted in 2010 maneuvered one hundred randomly selected satellites to match the orbits of another one hundred randomly selected satellites. Multiple runs of the simulation provided approximately the same results: within twenty-four hours from the commencement of hostilities, 95 percent of these simulated "attacks" were completed despite the fact that none were designed to minimize transit times or optimize start locations.* This means that adversary attacks must be quickly detected and their ultimate strategic goals assessed, in order to plan optimized countermaneuvering tactics as quickly as possible. In a word, flexibility is a problem. The military strategy, then, should focus on maximizing friendly force freedom of action in space while minimizing enemy freedom of action.

---

*   Paul Szymanski, Computer simulation of space attacks using Space Warfare Analysis Tools (SWAT) software he developed, 2010.

The third element, maneuver in the electromagnetic spectrum, remains the most elusive. Like the larger field of space operations, the subspecialty of electromagnetic spectrum operations (EMSO, in the doctrinal vernacular) has created a language based on the tasks to attack, defend, and support.* The ways and means of performing these functions vis-à-vis space warfare have yet to be fully developed, explored, or exploited. Because actions in the electromagnetic spectrum occur—with the proper planning and preparation—at the speed of light, they enjoy a significant potential for operational flexibility. To achieve such optionality, however, not only requires deliberate planning for the spectrum itself, but also deliberate planning for ground-segment weapons systems in a way that allows them to achieve surprise, confuse adversary response, and adapt to operational conditions without being overly predictable.

Offensive

Offensive—A principle of war by which a military force achieves decisive results by acting with initiative, employing fire and movement, and sustaining freedom of maneuver and action while causing an enemy to be reactive.
—Field Manual 101-5-1, Operational Terms and Graphics, 1997

Definitionally—at least by the formal, if eclipsed, doctrinal definition—the offensive links directly to the principle of maneuver. In the offensive, the attacking force seeks to gain its objective in such a way as to wrest from the enemy any possibility of employing its own plans to any effect.

When most successful, the enemy's activity—whether offensive or defensive—breaks down, its formation become incapable of organized activity, and a rout ensues. The rout becomes disaster if pursued by an organized force.

The propensity for organizational breakdown seems likely to be particularly pronounced in space warfare. Due to the remote nature of satellites

---

*   US Department of the Air Force, Annex 3-51, Electromagnetic Warfare and Electromagnetic Spectrum Operations, July 30, 2019, accessed September 20, 2020, https://www.doctrine.af.mil/Portals/61/documents/Annex_3-51/3-51-D01-EW-EMSOIntroduction.pdf.

in space, it is highly possible that a belligerent may initiate a small-scale space attack, execute the plan, and achieve his objective before the recipient even knows he is under attack, who is attacking, what their objectives or political aims are, or when political leadership can validate the attack and coordinate an appropriate response. Because of these factors, there is a good chance that the side who initiates space attacks will retain the initiative throughout their operation, particularly if they have capably set the time, place, and terms of the space battle, planned for tactical activities that will keep their adversary off balance with time-phased multiple attacks across multiple theaters with multiple types of weapon systems, and have a pre-approved plan to exploit successes for further gains. Even operational-level space attacks could feasibly be executed and completed within forty-eight hours, a danger compounded without adequate and timely space domain awareness, well-established military responses, and determined political will.

As such, space warfare plans should seek to seize, retain, and exploit the initiative, setting the time, place, and terms of the battle for the orbital, link, and ground segments. Depending on the desired political and military objectives and the political will to engage in such activities, this activity may include a preapproved ramp-up of hostilities to send a strategic message, to put the enemy off balance, or to set the conditions for exploiting future successes. Planners must consider aspects of tempo, including the speed of action, the ability to repeat desired activities, the amount of time necessary for coordination, and the potential benefits of synchronization across space missions and even across domains. They must also consider how their actions will affect the peace afterward, acknowledging the possibility that failure or even successes may cause realignment of allied relationships and new legal restrictions on space weapons.

In its treatment of maneuver, much of the preceding section dealt indirectly with the offensive, but what remains to be discussed is the space warfare relationship between the offensive and the defensive. The offensive, in the age of maneuver warfare, is dogma, but in practicality, the offensive and the defensive exist in tandem, one transitioning to the other as culmination—the point at which the offense or defense can no longer continue—occurs.* Furthermore, the offensive and the defensive exist in complementary fashion among domains and at the various levels of war. For example, since the early days of the Cold War, the United States has remained largely on the strategic defensive in space—albeit with periods of limited tactical offensives. Among other military action, this strategic defensive in space allowed for the strategic offensive in the form of the multidomain, intertheater movement of forces to set the conditions for Operation Desert Storm, characterized by its operational-level air and ground offensives. In truth, however, the air and ground offensives could not begin without operational-level air and ground defenses that set the conditions for the offensive (Operation Desert Shield).

As hinted at in the maneuver discussion, movement itself can be a kind of defensive action, especially for satellites. An adversary may have plentiful and superior space weapons, but if they cannot find or reach friendly space systems, their advantages are negated. Constantly moving so that an adversary cannot locate your satellite (continuous thrusting will confuse orbital dynamics software and sensor tracking by causing the spacecraft to deviate from the predictable rules of Keplerian orbital dynamics), your link segment traffic, or your ground segment operators; placing spacecraft in hard-to-see and hard-to-reach orbits (e.g., high inclination and highly eccentric orbits); building spacecraft, waveforms, and ground equipment with low-observable signatures; and even employing decoys as part of a deception plan will help ensure survivability in the face of aggression.

Nations with more military space systems naturally have more systems to defend, which may lead to an emphasis of defense over offense in technology development and in military planning. A belligerent with few space systems, in contrast, has fewer targets for offensive weapons. Thus, when facing such an enemy, an emphasis on the defensive seems generally prudent. With a high confidence in space domain awareness, including reliable knowledge of the enemy's (and its allies') offensive space weapons, however, it may be

---

*   Culminating point—The point at which a force no longer has the capability to continue its form of operations, offense or defense (JP 5-0). US Department of Defense, *DOD Dictionary of Military and Associated Terms* (Washington DC, 2020), 55.

possible to neutralize these systems early in the conflict before the enemy can fully implement his plans. This case again highlights the primacy of space domain awareness for conducting space warfare.

The implication of an emphasis on the defensive is that defensive capacity is finite. A country like the United States with a large number of assets is not likely to be able to defend all of its satellites, all of its link segments, all of its ground segments from all threats at all times.* If high demands are placed on defensive resources, the ability of a space power to provide protection for allies comes into question, and its ability to protect commercial entities seems even more tenuous—if that is indeed a military responsibility. This prognosis seems dire, and there may be a tendency to attempt to defend everything, but the space force and Space Command must resist this temptation. Instead, it must work to deliberately assess what assets must be defended for given operational scenarios and theaters—leaving the rest undefended if necessary.†

By way of historical analogy, the Allies of World War II could not effectively defend the entirety of the broad front. Prior to the Battle of the Bulge, much of the Ardennes region was under economy-of-force conditions, allowing the German offensive an opportunity to penetrate the Allied lines and make a drive for the Meuse River. In response, the Allies concentrated defenses at the most important road junctions, St. Vith and Bastogne. These defenses bought enough time for the Allies to organize a counterattack. There are two lessons for space operators in this scenario. First, not all positions or forces are of equal value in a given scenario. The most important must be defended vigorously, the rest economized or abandoned. Second, the defense sets the conditions for the offensive, which must be used to wrest back the initiative.

---

* As Frederick the Great said, "He who defends everything defends nothing."
† In *The Art of War*, Sun Tzu writes, "If he sends reinforcements everywhere, he will everywhere be weak."

Objective

> **Objective**—1. The clearly defined, decisive, and attainable goal toward which an operation is directed. 2. The specific goal of the action taken which is essential to the commander's plan. See also **target**. (JP 5-0)
>
> —DOD *Dictionary of Military and Associated Terms*, 2018

In the preceding section on the principle of the offensive, the road intersections at St. Vith and Bastogne during World War II became objectives for the Allies because of their importance to the success of future operations. Of course, in any operation, multiple objectives exist at the strategic, operational, and tactical levels. In planning, a consideration of the military-strategic objectives drives operational-level objectives, which drives tactical-level objectives. In execution, the process builds from the bottom up. Attainment of tactical objectives enables attainment of operational-level objectives, which contribute to the accomplishment of military-strategic objectives. Space warfare objectives primarily support terrestrial military operations, although a space war may erupt independently and have its own objectives.

As in other domains, military objectives may be terrain-based or enemy-based. For example, a tactical, enemy-based objective may be denial of enemy satellite observation over a battlefield for a given period of time. A "terrain-based" objective may be to defend a portion of the UHF spectrum in preparation for the main effort to achieve its objective. An additional "terrain-based" objective may be intensely defending the sun-synchronous orbits where imagery satellites tend to bunch up to achieve optimal lighting conditions when passing over the ground (an example of a space choke point). At the opposite pole, grand-strategic objectives may include preventing enemy access to key technologies or developing an information campaign to shape global opinion against the use of antisatellite weapons. If the objective is to eliminate a system capability or a category of information (for example, satellite imagery), the approach will be different than if the objective is to eliminate a given satellite or ground station. Such objectives, when linked geographically or by theme to military objectives in other domains, provide the stepping-stones for progress along lines of operation or lines of effort, respectively. As a prudent measure, of course, branches and sequels to space operations are necessary in case the initial efforts fail or in case objectives change.

Simplicity

> Simplicity—One of the nine principles of war. The preparation and execution of clear, uncomplicated, and concise orders and plans to facilitate mission execution in the stress, fatigue, and fog of war. See FM 100-5.
> —Field Manual 101-5-1, Operational Terms and Graphics, 1997

The principle of simplicity directs us once again to the question of a common language, the essential prerequisite for common understanding. A plan that is relatively simple to the practiced space operations expert may be nearly incomprehensible to practitioners of land, air, or sea warfare. Indeed, while soldiers and sailors may have difficulty coming to common terms on occasions, the domain of the one is at least somewhat intuitive to the other. After all, most soldiers have used guns, been on a boat, and are familiar with airplanes. A sailor may say that an aircraft carrier is like a floating city combined with an airfield, and the soldier could imagine it. Similarly, a sailor may have little firsthand knowledge of army operations but has likely seen movies about tank battles and parachute drops. Space operations, however, are not in the common experience of most military personnel—even though the military routinely relies on space-based capabilities.

An additional consideration in such discussions is the complexity of the space weapons themselves. While it is very easy to slip into technical discussions full of specialized jargon, the effects of space weapons must be easily understandable to friendly commanders. Likewise, the meaning of a space attack must be apparent to an enemy commander—a challenge if the enemy commander comes from a technologically challenged force or lacks understanding of the value of space to their overall military effort. If the desire is to elicit a response, an enemy must know that they have been hurt in the space domain. Depending on the adversary and their space domain awareness capabilities, it is quite possible that they may not know they have been attacked at all, or they may be unable to assess the overall effect on their operations. If surprise is desired, the adversary's ignorance is not a problem; if, however, the attack is calculated to send a message, some space domain awareness understanding among the enemy is required.

Preparing integrated plans—including branches and sequels—with other-service partners thus presents a particular challenge. Keeping the plan sufficiently simple is yet more difficult and often first requires some level of teaching members of other services just exactly what it is you are talking about. Warfighting frameworks like the principles, the elements of operational design, and military symbology are essential to relate space operations to the other more-established warfighting fields.

Restraint, Perseverance, and Legitimacy

The joint force added the final three principles of joint operations based on experience from "a variety of irregular warfare situations."* These terms do not have formal doctrinal definitions but follow common usage ones. All three require unique consideration in the discussion of space warfare, but among the three, restraint, with its common dual-usage throughout military doctrine, requires special critical thought.

In the joint planning lexicon, a restraint is something that must not be done. Depending upon the operation, there may be legal restraints, policy restraints, or restraints emplaced by the rules of engagement on how force may be used. As a principle, restraint takes on a more philosophical tone, intimating the need for commanders not only to understand the authorities imposed upon them by higher authorities but also to understand where restraint may be prudent within their own authorities. For space operations, the idea of restraint goes hand in hand with the desire to avoid orbital debris creation and conflict escalation. Whether restraints are imposed or adopted for the sake of prudence, the decision to not attack or to attack with nondestructive means may lead to long-term benefits. As Joint Publication 3-0 states, "reasons for the restraint often need to be understood by the

---

*   US Joint Staff, JP 3-0, *Joint Operations* (Washington DC: Government Printing Office, 2017), ix.

individual Service member, because a single act could cause adverse diplo-matic/political consequences."* This statement is especially true for space operations personnel whose tactical actions may have strategic consequences and who may be operating in support of multiple theaters simultaneously. The strategic consequences may even affect friendly forces. Denying an adversary's space capability, for example, may impact the intelligence col-lection abilities of other agencies who may be exploiting this target. It may even be prudent in some cases to attack "strange" space targets in order to confuse your adversaries and lead him down a predefined path.

Perseverance requires "patient, resolute, and persistent pursuit for as long as necessary."†Although conceived as a principle in response to the need for prolonged counterinsurgency operations, perseverance applies well as a principle for space operations because space operations exist in a global and continuous state. Because their work often spans the globe, because intelligence collection is never done, and because there is always a mission somewhere requiring space support, perseverance may be the most important principle for space operators. An infantry battalion, for example, will return home when the war is over. For space operators, the war never ends, and even routine operations require significant time. Moving satellites, whether to optimize communications coverage or to prepare for an antisatellite attack, can take months—timelines more on par with the relocation of seagoing vessels than aircraft. This condition requires patience, a predilection for long-term planning, a special amount of vigilance and focus on the part of the individual service members, and a deliberate effort to avoid complacency and to train while operating.

Finally, the principle of legitimacy implies the need to act in accordance with the guidance from established authorities, to avoid giving power to organizational entities that are not entitled to it, and to act in such a manner that would not undermine the goals of the mission.‡ To maintain legitimacy, commanders at all levels must be careful to not violate restraints

---

*   Ibid., VIII-27.
†   Ibid., VIII-28.
‡   Ibid.

and to not act in a way that commercial or international entities may view as irresponsible. The debris creation caused by the Chinese antisatellite missile demonstration of 2007 provides an example of a case in which a military and a nation lost legitimacy in the eyes of the world for irresponsible behavior. If war in space proves to be a more sensitive proposition to the general public than war on earth, space planners must take these sensitivities into consideration to make sure they do not violate international treaties and remain acceptable to political leadership. Because of the importance of international agreements and political acceptability, the Department of State should perhaps be consulted before initiating major space weapons program developments. Their perspective may enable early course corrections that may ultimately make such programs more acceptable on the international stage.

Conclusion

For Fuller, economy of force was more akin to a law than a principle. To his mind, the other principles were "so to speak, emanations of the one law as applied by our intelligence."* Economy of force involved not only the physical but also the mental and the moral spheres. In the attainment of an objective, all components were necessary, and a commander was responsible for marshalling them in the most judicious manner possible. Physically, commanders who enjoyed unity of command could mass forces, mental energy, and willpower on the most significant objectives while leaving smaller measures to achieve objectives of less significance. Maneuver, security, and surprise, when employed to effect, promised to lend a moral advantage, making up for lack of physical force.

In space warfare, the logic holds, although, as addressed, each principle comes with its own unique challenges and considerations. Massing is possible, but if massing in previous wars was accomplished through tens of thousands of soldiers, massing in space is accomplished with tens of satellites and in swaths of the electromagnetic spectrum. Due to a limited number

---

*   Fuller, *Foundations of the Science of War*, 220.

of space weapons systems and trained personnel, the rest must be economized. Surprise is equal threat and advantage; security is the outgrowth of the dread of surprise, but it is not generally a positive action. The defense, no matter how important, is not sufficient in itself and generally does not lead to military victory (although it may lead to the attainment of policy objectives). To seize the initiative requires the offensive and the attainment of progressive objectives.

As with all military theory, the principles of war remain good ideas until they cease to be or until an institution provides them a new meaning. Fuller himself, in his *The Foundations of the Science of War*, proposed the principles of direction, concentration, endurance, mobility, and others.* These have not come down to us in the dogmatic sense of the other principles but are worth a strong consideration, if not adoption. Conceivably, even the dogmatic ones may not always require implementation, but one would be wise to keep them in mind as a basis when employing other warfighting frameworks, including the elements of operational art and design.

No matter what the time or age in military warfare, many principles, rules, and "laws" of the past are applicable today and will continue to be so in future space warfare. Human intelligence, experience, training, culture, stamina, determination and aggressiveness all play a part in winning conflicts that machines cannot yet match. The personnel and war machines employed against opposing sides are merely "messengers" of intent and resolve that adversary commanders transmit to the opposing side to force their will upon the losing forces. Wars are still fought between adversary commanders at the strategic and operational levels and individual soldiers, sailors, marines, airmen, and space guardians at the tactical level.

---

* Ibid., 221.

CHAPTER 3

The Language of the Tactical Level

Section 1: An Overview of the Schema

As terms of art, *military strategic* and *operational level* are often debated among professionals, perhaps because their boundaries seem so difficult to map precisely. The tactical level of war, however, remains the most intuitive and therefore among the least controversial subjects. After all, most military professionals begin their training with the tactical, and it is quite possible for both officers and noncommissioned officers to spend thirty years in a service without ever venturing beyond the tactical level. Despite this familiarity, an intellectual framework, as in the discussion of the strategic level, is essential to further clarify the tactical level within the larger concept of space warfare and to set the scene for the ensuing discussion of tactical symbology, a necessity in the effort to develop a common language for the discipline.

Figure 12 depicts a conception of major activities within the tactical level of space warfare. It cannot be altogether comprehensive because the tactical level ultimately consists of all of the decisions of all of the individuals who are operating within it. However, as in other domains, the planning and execution of tactical operations must abide by the limitations of the operating environment, including its direct interaction with the operational level. It is important to note that Figure 12 is a schema, not a planning methodology. More formal planning methodologies like the joint planning process, the army's military decision-making process, the Marine Corps' planning process, and/or troop leading procedures still occur within this schema. Nonetheless, the schema allows for a discussion of some of the more salient activities of the tactical level of space warfare.

Broadly, the activities within the space tactical level follow the operations process: plan, prepare, execute, assess. They are, of course, directly linked to the operational level; the arrows on the left of Figure 12 indicate this relationship, but as discussed in the initial chapter on strategy, even the tactical level can influence the strategic.* The plans of the operational level drive the development of tactical-level plans—of which the chosen course of action (COA) will form a significant portion. The operational-level plan will include rules of engagement, and the tactical level will likely implement additional ones specific to its mission needs and possibly even to its proposed courses of action. Any tactical COAs will define criteria by which to measure success and a process of feedback to evaluate this success (or its degree of failure). The weaponeering and the sensor tasking will necessarily follow (bounded by legal, political and diplomatic considerations). If successfully executed, a course of action will lead to the termination of military space operations. Each of these components bears further discussion in turn.

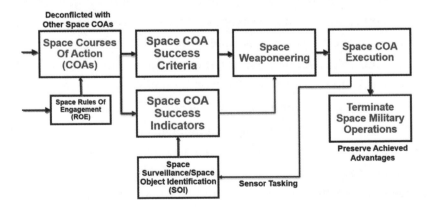

Figure 12: A Schema for the Tactical Level of Space Warfare. *Source:* Created by authors.

---

*   In the Operational Art chapter, we will argue that in contemporary warfare, the tactical level cannot lead to the accomplishment of strategic objectives without the operational level. The tactical level, however, can influence the strategic level indirectly and vice versa.

First, the question of rules of engagement for the tactical level is a delicate one because the employment of space warfare weapons could easily lead to destabilization or escalation in a conflict.* Short lead times for detecting and reacting to space attacks adds an additional complication, making the strategic space environment even less stable. A small, relatively inexpensive space mine, for example, could destroy a large, multibillion-dollar satellite that is crucial to the conduct of a nation's military operations. The action of employing the mine is a tactical one, but the ramifications of such a tactical action could very well have strategic consequences (conflict escalation, production of unacceptable amounts of orbital debris). As a result, the decision to employ space systems at the tactical level may not rest with the tactical commander.

The traditional response to this conundrum is for a senior commander to pull those employment decisions to his or her level. In one line of thinking, strategic-level decision-makers must make decisions of strategic consequence. This approach seems to have worked in the past because space has been a fairly benign operating environment, and the strategic-level decision-maker has not faced the danger of being overwhelmed by decisions. In future wars, however, it seems unlikely that any single commander will be able to address the myriad of decisions required across the multiple different space operations missions within the surface, electromagnetic, and orbital segments of the space domain. Part of the solution to this problem is allowing the appropriate decision authorities at the appropriate echelons for the appropriate missions. Another part of the solution lies in planning, with clearly defined rules of engagement and decision criteria. Perhaps the largest part of the solution, however, lies in building a force that is trained and educated to understand not just tactics but the holistic enterprise of space warfare. As discussed in the opening chapter, building the force is properly a military strategic problem with both institutional and warfighting equities across all levels of war. For the tactical level, planning will proceed as guided by higher rules

---

*   For discussion on the space conflict escalation ladder, see Chapter 4: "The Language of the Operational Level."

of engagement and established norms of behavior, including delegation of authorities, which will feed the development of courses of action.

The term *course of action*, like the term *strategy*, is one that is often used with imprecision. Doctrinally, a course of action is "1. Any sequence of activities that an individual or unit may follow or 2. A scheme developed to accomplish a mission."* This definition rings with the colloquial use of "strategy," but a course of action is not a strategy, nor is it a plan. A COA is part of the planning process—"a broad potential solution to an identified problem"†—that, when complete, drives additional detailed planning that is ultimately manifested in a written order. Furthermore, within the tactical level, COAs typically only apply to echelons with staffs (those commanded by lieutenant colonels and above) while tactical echelons without staffs—due to time and resource constraints and their nature as executors—develop singular plans to accomplish their assigned mission. They thus rely on their higher headquarters for the thorough analysis and much of the critical thought about the bigger picture.

As military planners quip, a plan is never finished, only abandoned, and in truth, there is no such thing as a perfect plan. There will always be gaps in information, oversights, misinterpretation, and changes in the surrounding environment, which is why plans change and new orders replace the older ones. While the completeness of a strategy may be debatable, the joint publication on planning defines the criteria for completeness of a military plan:

Completeness—"The plan review criterion for assessing whether operation plans incorporate major operations and tasks to be accomplished and to what degree they include forces required, deployment concept, employment concept, sustainment concept, time estimates for achieving objectives, description of the end state, mission success criteria, and mission termination criteria." (JP 5-0)

This definition lends itself to joint planning at higher echelons—particularly the development of contingency plans at combatant commands—yet

---

* US Department of Defense, *DOD Dictionary of Military and Associated Terms* (Washington DC, 2020), 53.

† US Department of the Army, Field Manual (FM) 6-0, *Commander and Staff Organization and Operations* (Washington DC: Government Printing Office, 2014), 9–16.

it remains useful for space forces at all echelons. It is, in truth, an ends-ways-means framework, suggesting the ubiquity of that concept, even at the tactical level. Tactical space units that often task, organize, and deploy as independent detachments or teams must strive to understand the completeness of the plans to which they are a part because they integrate with other organizations, likely receiving sustainment and protection from a host unit as they work toward the accomplishment of their objectives. An appreciation for the entire plan provides one with a greater understanding of how one might better contribute to it.

Within the planning process nest COAs and COAs themselves have criteria for what make them complete. By army doctrine, they must include the following:

- "how the decisive operation leads to mission accomplishment
- how shaping operations create and preserve conditions for the success of the decisive operation or effort
- how sustaining operations enable shaping and decisive operations or efforts
- how to account for offensive, defensive, and stability or defense support of civil authorities' tasks
- tasks to be performed and conditions to be achieved"*

In another view, a complete COA includes a concept for each warfighting function: command and control, maneuver, fires, intelligence, sustainment, and protection.† Rather than stand-alone concepts, each of these concepts should be integrated with the others, and each should contain as much flexibility as possible. If, for example, the sustainment plan relies on a single road for resupply, it is not flexible. If a command-and-control scheme relies on a single type of satellite communications, it lacks flexibility.

---

*   Ibid., 9–17.

†   While the Joint Force refers to these as the "joint functions," the authors here choose the army term "warfighting functions," judging it the more appropriate term to apply to space as a warfighting domain.

From a critical examination of the courses of action, success criteria emerge. Success criteria are those conditions that should be met to increase the likelihood of the operation achieving its desired outcomes. It is possible that an operation may achieve its desired outcomes without achieving all predefined success criteria, but in general, the success criteria provide guideposts and may even serve as intermediate military objectives.

Success indicators perform a function similar to success criteria, but they are less conclusive, often only hinting at the possibility of action. If success criteria are guideposts along the road, indicators are a blaze along the trail—less obvious but noticeable to those who are properly trained to recognize their significance. Indicators may relate to success criteria (in a case where multiple indicators provide sufficient evidence to determine that one or more success criteria have been met), or they may stand apart (in a case where the evolution of the operation has made obsolescent predefined success criteria, but indications from intelligence sources combine to suggest the intentions of the enemy's actions). Among the uses of success criteria and success indicators are in weaponeering efforts.

Weaponeering is the process of aligning available weapons to designated targets, the effects on which will contribute to the attainment of tactical-level, operational-level, and possibly even strategic-level objectives. In a traditional air campaign, for example, weaponeering involves matching a target (enemy bunker) to a munition (500-pound bomb) to a platform (attack aircraft) and then backward planning to sequence aircraft availability and logistical support. For space warfare, the process is conceptually the same, but the means vary considerably.

A space conflict, just like an air conflict, involves multiple warfighting domains. An air war requires ground basing, overland and oversea resupply, multiple types of air activity (reconnaissance, defensive, and offensive measures), and access to the electromagnetic spectrum. A space conflict requires similar considerations and employs means on land, air, and sea through the electromagnetic spectrum and in space to successfully execute a course of action, either independently or as part of a larger joint operation.

So, what might weaponeering look like for space operations? First, one must consider the orbital aspect. If we assume that the threat capabilities envisioned by the Defense Intelligence Agency are feasible in the near future, then we may begin to imagine some possible scenarios. It may be necessary, for example, to employ an on-orbit jammer/cyber weapon against an enemy communications satellite. The payload (in this case, the jammer) may be one of many payloads on the same spacecraft. If another payload is for relaying point-to-point ground communications, the weaponeering must consider the amount of power available to the spacecraft, the tradeoffs between using alternate payloads, the time required to reach the target, the effective range of the jammer, and the amount of fuel necessary to maneuver. It is further possible that only one channel on the target satellite requires negation, and other shared resources on the satellite—perhaps in use by neutral nations or commercial entities—must be taken into consideration. One must furthermore consider how to assess the effectiveness of the weapon, the extent of any collateral damage, and what options are available to both the defender and the attacker if the original attack fails.

A similar process occurs with the weaponeering of ground-based weapons. Whether a direct-ascent satellite missile, a ground-based jammer/cyber weapon, or a directed energy/laser weapon, all require a robust intelligence network, including space surveillance data, information about enemy communications signals, an understanding of how the target fits into the enemy's operational intent and scheme of maneuver, and the ability to assess the effectiveness of the weapon's employment. This assessment, like the assessment of on-orbit activities, provides indicators of success (or failure) and will allow the commander and staff to assess the success of the course of action. Although not depicted in the preceding schema, an unsuccessful activity should rightly cause the command to return to the drawing board—maybe revisiting the feasibility of their entire course of action or, in a less dramatic case, revisiting their weaponeering solution as part of their course of action and continuing to execute. One must always have a plan "B" or even plan "C" in place in case of failure of plan "A."

A third aspect of space weaponeering is the weaponeering that does not involve "space" weapons at all. An operation may require, for example, a rocket attack or special forces attack against an enemy satellite control ground station, which requires weaponeering. Alternatively, rather than employing artillery, a friendly cyber force may negate an enemy's ability to transfer information throughout its computer networks. These capabilities, just like orbital or ground-based jammers, require weaponeering and pos-temployment assessment.

As vital as weaponeering is to the execution of any military operation, it is important to remember that tactics are more than just targeting. We stress the idea here because considering what weaponeering looks like from a space operations perspective is very new to much of the joint force. Tactics involve the orderly movement and employment of forces in the face of the enemy, and although the idea of weaponeering only considers offensive capabilities, space operations (and any military operation, for that matter) require a consideration of all defensive capabilities, as well.

There are additional concerns for space weaponeering, such as timing and tempo, and combined arms attacks (sequencing terrestrial attacks with space attacks), all within the bounds of rules of engagement and policy considerations. The possibility of long-term effects of space attacks are very real, including the residual effects of space debris generation, the realignment of allies, and possible international condemnation and restrictive space treaties. Maybe a country can "win" the space war but lose the resulting peace afterward. Weaponeering planners should also take into account the subsequent viability of space weapon systems after they have been revealed to adversaries through attacks that can easily by countered the next time around. The degree and types of space attacks can shape current and future space weapons development programs and attitudes that are potentially detrimental to friendly space conflict options and to the world order in general.

This broad overview of the tactical level of space operations can in no way be conclusive; tactics are simply too dynamic, too greatly influenced by a myriad of factors inside of and outside of the operating environment. The schema does, however, provide a useful starting point from which to begin applying other frameworks, from which to continue to build the language of tactics, allowing practitioners to assemble the various pieces of the puzzle together in their own minds, making sense of and applying what is useful and possibly even discarding—after careful consideration, of course—what is not useful in a particular instance. To continue a discussion on the language of tactics, we must address in detail the need for a language of military symbology—a fundamentally tactical language because it breaks down space warfare into its constituent parts: the individual unit or spacecraft, the singular action, and the graphical-spatial means to regulate their behaviors. This language is fundamental to the execution of the tactical schema previously outlined and a common understanding across the levels of war.

Fortunately, the US military's robust system of symbology provides an efficient linguistic approach for planning and graphics development, an agreed-upon means for visualizing current operations, and a powerful aid in achieving situational understanding. In short, military symbology visually represents the common ideas and concepts (for example, doctrinal terms) that form the basis of the joint force's common language. As robust as the system of symbols is, however, there are significant barriers to its utility in military space operations. While some symbology does exist specifically for space operations, the available lexicon is not sufficient primarily for three reasons. First, the available symbology does not take into account space as an evolving warfighting domain or space operations as inherently multidomain operations. Second, even where the lexicon is robust, the space portions of it are not, at this time, well institutionalized across the various services. Third, currently existing space icons do not even cover all of the types of satellites already in orbit, much less future planned space systems. Because the common language necessary for planning, visualizing, and understanding space operations is incomplete, space operations themselves are, as a consequence, deprived of an opportunity for increased effectiveness. In addition, commonality in language and military concepts between traditional terrestrial military operations and those of space help better integrate space into current military operations. If nonspace military personnel can easily understand the space situation simply by viewing a common military visual display map that looks 90 percent like a traditional terrestrial military situation map, we have come a long way to integrating the two disparate types of operations.

In practical terms, a common tactical symbology enables improved tactics, the combination of tactical actions brings about an operational-level application of space operations, and success at the operational level achieves—wholly or partly—strategic military objectives. Fortunately, the available domain-specific symbology (space, ground, air, maritime) and functional symbology (for example, symbology of the intelligence function) provide the baseline for achieving a working symbolic language for space operations. Operational needs require the creation of additional symbols by combining existing icons, modifiers, and amplifiers or by simply creating new ones. With this baseline, the joint force may begin building a common understanding of available symbols for space operations that are based on a partially shared language. Furthermore, the pairing of unit symbols to tactical task symbols and graphic control measures suggests how those symbols may combine for coordinated tactical actions across multiple domains (more in the next chapter). With a common symbology established, space operations may become a more effective contributor to all-domain operational art (the topic of the next chapter).

Section 2: Spacecraft and Unit Symbols

Useful symbology comes from multiple sources, including joint and service publications. A discussion of joint force symbology necessarily begins with *The Department of Defense Interface Standard on Joint Military Symbology.** This publication focuses on "the display of symbols in modern multichromatic electronic systems."† In other words, it is a technical publication (rather than a tactical one) that aims to standardize symbology for use in common operating pictures. Color and shape schemes for symbols of all domains follow in Table 1. As in other domains, friendly space symbols are blue and rounded, enemy symbols are red and invoke the diamond shape, neutrals are green and square, and unknowns are yellow with a

---

*   US Department of Defense, MIL-STD-2525D, *Department of Defense Interface Standard*, 2014.
†   Ibid., 8.

trefoil or quatrefoil shape.* A gray (rather than a black) outline indicates that the icon's status is assumed (e.g., "assumed friend" or, if an enemy, a "suspect").† An icon representing a civilian entity may show a lavender fill with its relationship status (friendly, neutral, or unknown) following the icon shape convention (round, square, or trefoil, respectively).‡

DIMENSION / STANDARD IDENTITY	UNKNOWN	SPACE	AIR	LAND UNIT	LAND EQUIPMENT AND SEA SURFACE	LAND INSTALLATION	SUBSURFACE	ACTIVITY/ EVENT
PENDING (YELLOW)								
UNKNOWN (YELLOW)								
FRIEND (CYAN)								
NEUTRAL (GREEN)								
HOSTILE (RED)								
ASSUMED FRIEND (CYAN)								
SUSPECT (RED)								

Table 1: Frames Depicting Standard Identities and Dimensions. Source: MIL-STD-2525D, p. 12.

Importantly, MIL-STD-2525D gives symbolic options for tracking the health and status of individual icons (Table 2). A black slash across the icon indicates the entity has been "damaged/rendered ineffective," while a black "X" across the icon indicates destruction.§

---

*   Ibid., 13.
†   Ibid.
‡   Ibid., 17.
§   Ibid., 31.

DIMENSION / OPER. CONDITION	AIR/SPACE	SURFACE			SEA SURFACE	SUBSURFACE
		UNITS	EQUIPMENT	INSTALLATIONS		
FULLY CAPABLE	(icon)	N/A	(icon)	(icon)	(icon)	(icon)
DAMAGED/RENDERED INEFFECTIVE[1]	(icon)	N/A	(icon)	(icon)	(icon)	(icon)
DESTROYED	(icon)	N/A	(icon)	(icon)	(icon)	(icon)

Table 2: Operational Condition Amplifiers for Icon-based Symbols. Source: MIL-STD-2525D, p. 31.

In an alternative depiction (Table 3), a green bar beneath the main icon indicates a fully capable entity, a yellow bar is equivalent to a slash, and a red bar is equivalent to an "X."* A blue bar beneath the main icon indicates that the entity is "full to capacity." Although the MIL-STD does not suggest a specific case for its use, in space operations, a communications satellite may be assessed as "full to capacity" for a given amount of time, if it cannot support additional users. To take the thought one step further, a communications satellite with multiple payloads, for example, could employ multiple bars to denote the capacity status of each type of payload.

---

*   Ibid., 32.

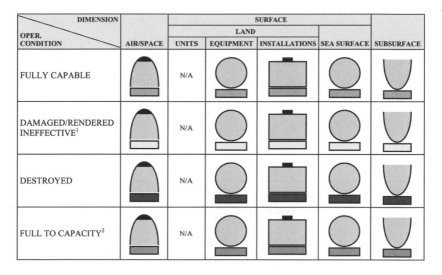

DIMENSION / OPER. CONDITION	AIR/SPACE	SURFACE				
		LAND			SEA SURFACE	SUBSURFACE
		UNITS	EQUIPMENT	INSTALLATIONS		
FULLY CAPABLE		N/A				
DAMAGED/RENDERED INEFFECTIVE[1]		N/A				
DESTROYED		N/A				
FULL TO CAPACITY[2]		N/A				

Table 3: Alternate Operational Condition Amplifiers. Source: MIL-STD-2525D, p. 32.

In addition to pictorial amplifiers, icons may also include text amplifiers. Figure 13 shows the current standardized way of depicting space symbol amplifiers. These amplifiers include (from top to bottom) information on the satellite's engagement status, its track number, the satellite type, its speed (not a useful parameter for orbital warfare), a portion for additional comments, and at the bottom, the operational condition bar (as illustrated in Table 3). This method, however, provides only part of the information necessary for dynamic military decision-making.

Figure 13: Current Standardized Way of Depicting Space
Symbol Amplifiers. *Source:* Created by authors.

The effectiveness of future space operations requires a more complete method for depicting amplifiers, including the information required not only to envision the spacecraft's current state but to more deeply appreciate its potential for future operations. Although a more complicated method of annotation, the method of employing amplifiers shown in Figure 14 is necessary to account for the complexity of the orbital environment where thousands of objects belonging to hundreds of actors are in constant motion—either from the force of gravity or because of deliberate maneuvering.

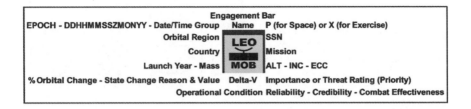

Figure 14: Proposed Way of Depicting Space Symbol Amplifiers. *Source:* Created by Paul Szymanski.

The new method retains the engagement bar at the top with the satellite's name underneath. Not only will the satellite's name provide insight into its function (especially when part of a constellation), but the name provides an important reference point from which to begin deeper analysis into existing intelligence information. On the left side of the icon, the top row shows the time data of the most recent information obtained about the satellite. As with all intelligence collection activities, the information about a satellite only remains useful for so long after being collected (orbital elements are typically good for one to two weeks). Furthermore, because existing collection methods cannot maintain current information on every satellite at all times, it is important to know the last time the satellite was observed. If the last collect is from several days ago, any actions may first require verification of the satellite's current state and an accurate location.

The next line below the time information denotes the orbital region. This designation is intended to be more specific than just the common orbital regimes (for example, low earth orbit). As the section on divvying up the space area of responsibility (AOR) will discuss, it is possible to designate specific regions within space—whether these are joint operations areas, space defense identification zones, or some other defined subregion.

Information about the country of origin follows. Like Figure 13, Figure 14 employs a green rectangular icon, indicating that the satellite belongs to a neutral nation, but depending upon the current geopolitical situation, one neutral nation may be on the verge of becoming an adversarial nation. Similarly, a nation-state may be an acknowledged neutral in the grand strategic sense, but its behavior in the space domain may not fall within desirable norms and therefore requires watching. In the cases of satellites that belong to international corporations, it will be helpful to know the country of origin. In a conflict, a commercial satellite may be considered neutral, even if its primary stakeholder is an adversarial nation. Alternatively, depending upon the rules of engagement, a commercial satellite may be targetable, if it is aiding the enemy's war effort and if such targeting falls within the rules of engagement, the laws of armed conflict, and the commander's intent.

The line below the country-of-origin accounts for the launch year of the satellite and its mass. The age of the satellite provides an indicator of its expendability. If it was launched twenty years ago, which is not uncommon, it may be near the end of its usable life and is not likely to have advanced technology on board. In a conflict, however, this satellite may be one of the first that the enemy is willing to sacrifice by forcing it to collide with a more valuable friendly asset. Like the launch date, the mass of the satellite provides information about its potential function and its potential danger. A large satellite may be roughly the mass of a school bus while the smallest satellites could fit in the palm of one's hand. Even a small satellite could damage a much larger one, and because of their size, these provide particular difficulties for collection assets. Amplifiers to provide more precise information about the size of satellites are proposed in Table 6.

Returning to the top and working down the right side of the icon, a "P" will indicate a normal operational icon where the "X" would denote the icon's role as part of an exercise scenario—as in the legacy method. The Space Surveillance Network's (SSN) numerical designation follows, which is useful for cross-referencing, database searching, and clarity in requesting support or conveying operational facts. There are, for example, many Russian COSMOS satellites, but each one has a unique number designation.

The line for amplifying mission information also helps to bring home the uniqueness of the spacecraft in question. It is true that the symbol itself provides much information about the satellite's mission. In Figure 14, for example, the icon shows that the satellite is a communications satellite in low-earth orbit that is serving mobile users (indicated by the modifier "MOB"). By itself, one may infer that the satellite is supporting a surface user, but additional information about the mission—such as "Supporting Fifth Fleet"—can amplify the operational significance of the spacecraft.

If the mission line provides literal information about what the satellite is doing, then the altitude, inclination, and eccentricity of the satellite's orbit suggest what it might do. As will be discussed in the chapter on the operational level of warfare, these parameters are of highest importance when considering the possibility of orbital intercepts. The information contained in the row beneath the symbol contributes to this understanding. At the left is information about the percentage of orbital change, the reason for the state change and other satellite characteristics, and an associated state change number. Higher state change numbers may be colored red, providing an additional visual cue. If, for example, four of the highest state changes belong to the Chinese BeiDou constellation, it means that they have significantly altered the orbital parameters, size (for example, by jettisoning a subsatellite), or attitude of those satellites since the last observation, which may indicate that China is optimizing its constellation for a specific purpose. The "delta-v," the ability of a satellite to change its velocity, is a rough proxy for the amount of fuel on board and provides insight into how much it could maneuver (more on maneuver envelopes later).

The next piece of information is the importance rating (friendly or neutral spacecraft) or the threat rating (enemy spacecraft). These ratings put a number to the spacecraft, prioritizing assets to either collect intelligence against the threat or to protect friendly assets from the many threats. Lower threat numbers (for example, the highest threat, #1) may be shown in red while the highest priority friendly satellite rankings may be shown in blue. The last item on this row at the far right is information on the rules of engagement (ROE). Although a visual method of depicting ROE is developed in the section on spacecraft symbology, an alternate method includes a two-letter designation for the various types of engagement allowed against an enemy. Figure 15 lists these designations.

<u>ROE Ratings (Maximum Allowed Attack)</u>:
D1 Deception
D2 Disruption
D3 Denial
D4 Degradation
D5 Destruction

Figure 15: Codes within the ROE Rating Scheme. *Source:* Created by Paul Szymanski.

Finally, at the very bottom of the symbol rests information on the operational condition evaluating rating. Because intelligence information contains uncertainty, the reliability and credibility ratings provide information on the veracity of the information. For example, a satellite with a reliability rating of "A" and a credibility rating of "1" indicates that the information is judged as "completely reliable" and "confirmed by other sources." The combat effectiveness is likewise subjective and amplifies the color bar at the bottom of the symbol, if one is employed. A satellite's combat effectiveness rating may range from fully operational ("FO") to not operational ("NO") with an additional modifier to indicate that combat effectiveness is unknown ("UN"). Figure 16 lays out the codes within the reliability, credibility, and combat effectiveness rating schemes.

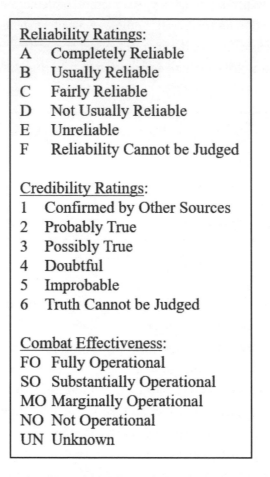

Figure 16: Codes within the Reliability, Credibility, and Combat Effectiveness
Rating Schemes. *Source:* Created by Paul Szymanski.

Adding an arrow to the base frame shown in Figure 14 can provide
further information about the satellite. To denote a change in location, one
may use a downward arrow indicating a decrease in altitude, an upward
arrow indicting an increase in altitude, and a left- or right-pointing arrow
indicting a drift westward or eastward, respectively. Figure 17 shows an
example of how such a state change symbol may look. Note that in this
example, the downward arrow also has the words "Maneuver, Decrease
Altitude" superimposed over it.

Figure 17: State Change Arrow Indicating Decrease in Altitude. *Source:* Created by Paul Szymanski.

Several additional arrows are useful. For example, Figure 18 uses an arrow pointing to the eleven o'clock position to indicate a satellite's inclination change. The arrow is red because the intent of the change is unknown. Figure 19 shows an arrow pointing to the seven o'clock position to indicate the decay and breakup of the satellite. This arrow is also red because it poses a threat, in this case to South Asia.

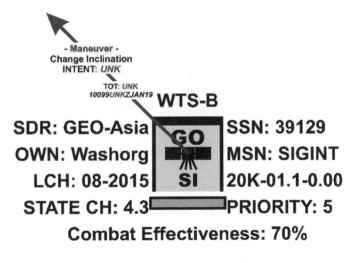

Figure 18: An Arrow Pointing to the Eleven O'Clock Position Indicates an Inclination Change. *Source:* Created by Paul Szymanski.

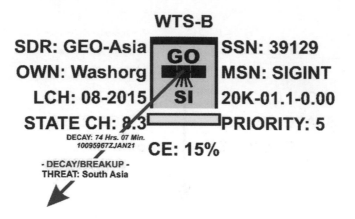

Figure 19: An Arrow Pointing to the Seven O'Clock Position Indicates Decay/Breakup. *Source:* Created by Paul Szymanski.

Similarly, an arrow pointing to the one o'clock position indicates an antisatellite (ASAT) attack (Figure 20), and an arrow pointing to the five o'clock position indicates that the satellite intends a rendezvous and proximity operation (RPO, Figure 21).

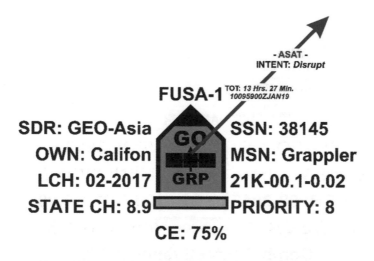

Figure 20: An Arrow Pointing to the One O'Clock Position Indicates an ASAT Weapon Attack. *Source:* Created by Paul Szymanski.

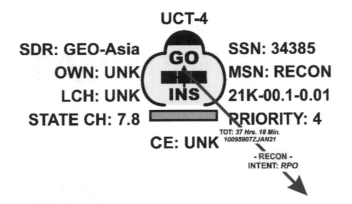

Figure 21: An Arrow Pointing to the Five O'Clock Position Indicates an RPO Operation. *Source:* Created by Paul Szymanski.

With the basics of symbology in hand and with a system for providing amplifying information about them, it is now possible to explore the wide variety of symbols that may be employed in space operations. These symbols, of course, represent equipment, units, and ways of employing them—the building blocks of tactics and the means of visually communicating battlefield action. Just as a junior army officer would begin by learning the symbols for individual weapons systems like tanks and machine guns, we continue our discussion with a variety of symbols for on-orbit means, including many unaccounted for in the current symbolic lexicon.

On-Orbit Means—General

With general symbolic conventions addressed in the main body of MIL-STD-2525D, Appendix B of that document deals specifically with symbols for on-orbit capabilities. Many of the icons listed in the appendix are not likely to be useful on a routine basis (for example, the planet lander or the biosatellite symbols). Table 4, however, shows the most practical icons and modifiers. As with all symbols, the icon fits within the frame and may have modifiers above and below it.

Row	Description	Icon
1	Satellite, General	
2	Satellite	
3	Antisatellite Weapon	
4	Communications Satellite	
5	Earth Observation Satellite	

6	Miniaturized Satellite	
7	Navigation Satellite	
8	Reconnaissance Satellite	
9	Weather Satellite	
10	Space Launch Vehicle	

Row	Description	Modifier
11	Orbiter Shuttle	
12	Capsule	
13	Space Station	
Row	**Description**	**Modifier**
11	Low Earth Orbit (LEO)	
12	Medium Earth Orbit (MEO)	

13	Highly Elliptical Orbit (HEO)	
14	Geosynchronous Orbit (GSO)	
15	Geostationary Orbit (GO)	
16	Molniya Orbit (MO)	
17	Optical Sensor	

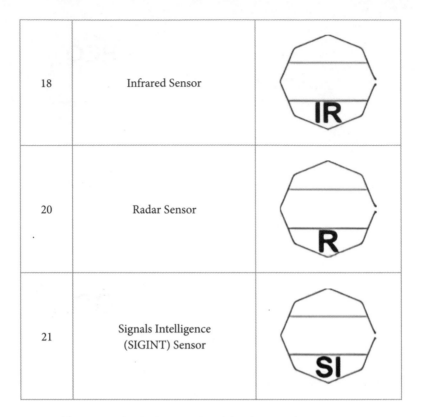

18	Infrared Sensor	
20	Radar Sensor	
21	Signals Intelligence (SIGINT) Sensor	

Table 4: Practical Orbital Icons and Modifiers for On-Orbit Assets. Source:
Table compiled by authors based on MIL-STD-2525D, Appendix B.

Upon examination of these symbols, several recommendations emerge. First, the "Satellite, General" graphic shown in Row 1 is redundant to the "Satellite" graphic shown in Row 2. While having multiple means of expressing the same capability is common practice in military symbology, Row 2's satellite may be the more useful of the two, invoking as it does a visualization of a satellite. Row 3's "Antisatellite Weapon" is nonspecific, and given anticipated types of antisatellite weapons, this base icon will require additional modifiers; the section on orbital warfare symbology takes up this discussion. Row 4's "Communications Satellite" is similarly generic and requires additional modifiers to specify the type of communications satellite. For the purposes of tactical symbols, the modifiers proposed in Table 5 are likely to be sufficient.

Row	Description	Proposed Modifier
1	Ultra-High Frequency Communications Satellite	UHF
2	Super-High Frequency Communications Satellite	SHF
3	Extremely High Frequency Communications Satellite	EHF
4	L-Band Frequency	LBD
5	S-Band Frequency	SBD
6	C-Band Frequency	CBD
7	X-Band Frequency	XBD
8	Ku-Band Frequency	KUBD
9	K-Band Frequency	KBD
10	Ka-Band Frequency	KABD

Table 5: Proposed Modifiers for the Communications Satellite Icon. Source: Created by Jerry Drew.

In a similar way, the reality of "Miniaturized Satellites" (Row 6) has surpassed the symbolic language available to express them. With satellites of ever-decreasing sizes, new terms have been coined for the sake of precision. It is quite possible that some very small satellites are not currently trackable with existing means, but it is also possible that they may be in the future. It is also possible that multiple small satellites (of whatever flavor) could be employed in swarm configurations. This level of specificity belongs more within the subcategory of amplifiers (vice icons or modifiers). The proposed amplifiers in Table 6 adopt real-world terminology, but their applicability to military operations is admittedly forward-looking.

Row	Description	Proposed Amplifiers
1	Minisatellite (100 to 500 kg)	<500kg
2	Microsatellite (10 to 100 kg)	<100kg
3	Nanosatellite (1 to 10 kg)	<10kg
4	Picosatellite (100 g to 1 kg)	<1kg
5	Femtosatellite (10 to 100 g)	<100g
6	Attosatellites (1 to 10 g)	<10g
7	Zeptosatellites (0.1 to 1 g)	<1g

Table 6: Proposed Amplifiers for the Miniaturized Satellite Icon. Source: Created by Jerry Drew.

Finally, Row 9's "Weather satellite" combines the satellite icon of Row 2 with the letters "WX" as part of the icon. To remain consistent with other symbology, the letters "WX" should become an acknowledged modifier for the base satellite icon rather than remain part of the icon itself.

On–Orbit Means—Not Yet Accounted For

The rest of the existing symbology in Table 4 bears little comment, but noticeably, a large number of space-based capabilities and objects are absent from the lexicon (not including the anticipated weaponized satellite types discussed in the next section). Some types of objects, like missile warning satellites and space debris, have existed for decades, but there is not adequate symbology to depict them. Other capabilities, like video imagery from space and space domain awareness satellites, are newer, but established symbology again faces a shortcoming. Icons for the major types of capabilities fill part of the symbology gap. The ability to account for the wide variety of payloads through modifiers fulfills the second part of this gap. Niche capabilities, like scientific or dummy payloads, may not need their own icon, but accounting for their presence on orbit requires the creation of unique modifiers to be added to the icons. Table 7 shows proposed icons followed by proposed modifiers. Note that some capabilities may be represented as either icon or as a modifier paired with a more generic icon (a generic satellite icon, for example). This flexibility occurs in the symbology of other domains and is common practice.

Missile Warning Satellites (Row 1, Row 7)

Despite the long-established importance of missile warning satellites to both strategic and tactical warning, no dedicated symbol for a missile warning satellite exists. A combination of either the "earth observation" or the "reconnaissance" icon with an "IR" modifier may hint at the function of overhead persistent infrared (OPIR) surveillance, but neither combination is sufficiently clear. As a modifier, "MW" already refers to "mine warfare" in another part of the symbolic lexicon, and it may be unwise to invite confusion on this point; it is therefore not available for "missile warning." Furthermore, since reconnaissance is conceptually different than surveillance, use of the "reconnaissance" modifier does not seem appropriate. The most direct way of depicting a missile warning satellite, then, may be through employment of the base satellite icon with a new modifier for overhead persistent infrared ("OPIR"). This symbol/modifier combination is shown in Table 7, Row 1. "OPIR" may also be used as a modifier with another satellite icon, as shown in Row 7.

Space Domain Awareness Satellites (Row 2, Row 8)

Like missile warning satellites, situational awareness satellites do not fit neatly into the "reconnaissance" label. They may perform reconnaissance of an area or of another spacecraft, but they may also perform surveillance, an activity akin to reconnaissance but requiring significantly longer duration. Thus, the reconnaissance icon may be misleading. To distinguish general situational awareness from space domain awareness (SDA), the recommended icon for a space domain awareness satellite is the letters "SDA" (Row 2). As with "OPIR," it is possible for "SDA" to serve as a modifier with another icon (Row 8).

Reentry Vehicles (Row 3, Row 9)

As shown in Table 4, symbols exist for capsules and orbiters, both of which are designed to reenter the atmosphere. An icon does not exist, however, for a generic reentry vehicle, which may be useful in designating jettisoned payloads of interest (as in the case of the historic CORONA program) or independent warheads released by an exoatmospheric missile (like the US Minuteman or the Russian SS-19). For a satellite returning to earth, designation of the satellite as a reentry vehicle may be helpful in tracking its status and in distinguishing it from the numerous other trackable satellites (modifier variation in Row 9).

Video Imagery Satellites (Row 4, Row 10)

Another capability that requires modification of extant symbology is the video imagery satellite. The commercial company Terra Bella (formerly

SkyBox Imaging) successfully demonstrated this technology in 2013, and it is reasonable to assume that both state and nonstate actors will leverage similar technology in the future. The army employs a "Video Imagery (Combat Camera)" icon (Row 3) to depict Signal Corps units who serve as battlefield videographers. Although a satellite with video capability performs more of a reconnaissance than a videography function, the video camera icon itself is a natural choice to pair with a space unit frame. Alternatively, the icon could be adapted into a modifier for use with the generic satellite icon (Table 7, Row 10).

Debris (Row 5)

Without a way of depicting orbital debris, space operations forces cannot have a complete picture of the operating area. Orbital debris is so prevalent and such a threat to existing satellites that considering its impacts is essential in any space mission planning. There is, of course, a limit to how useful such an icon can be; it would not be feasible or practical to track every piece of space debris. As the pictograph in Row 5 suggests, employment of the icon may be most useful when considering large debris clouds, particularly as they may diverge after a collision and threaten nearby assets. There is also the consideration that threat satellites might pose as debris in order to hide their true intents.

Refueling Satellites (Row 6, Row 11)

Refueling satellites do not currently exist on orbit. However, several experimental ventures have explored the possibility, and it seems only a matter of time—probably a short time—before purpose-built satellites are refueling their neighbors. Refueling satellites perform a function similar to tanker aircraft in the air force. In air domain symbology, the icon for a tanker is a capital "K." When used with a space-domain frame, the combination indicates a refueling satellite (Row 6).

Alternatively—and in keeping with the optionality of the air domain usage—the "K" can serve as a modifier for the satellite icon (Row 11).

Because the most important factor in military space warfare is the ability to maneuver, future space conflicts may see the "topping off" of critical constellations with fuel before initiating operations—a possible indicator of conflict escalation.

Additional Modifiers

Future capabilities aside, the wide variety of satellites and payloads already on orbit requires the implementation of new modifiers to account for civil and commercial uses in themselves but also to consider the depiction of a satellite with dual military and commercial uses. This could be either a military satellite hosting a commercial payload or a commercial satellite hosting a military payload. The benefits of such an arrangement include cost-sharing, deception, and deterrence. The Department of Defense, for example, may not wish to pay for an entire satellite and incur the long procurement time lines if it can fit a small, readily available payload on a commercial satellite that will do the job. Since this small payload is on a commercial satellite, it may go largely unnoticed by adversaries, providing a capability that does not draw much attention to itself but could serve a vital backup function if similar capabilities were lost. Furthermore, if the enemy knows that there is a defense payload on a commercial satellite, it may not wish to attack a commercial asset—particularly if such an attack would damage the economic interests of their allies or neutral parties. To depict such a joint military-commercial satellite, the modifier "JON" is proposed (Table 7, Row 12).

Of the variety of capabilities that could serve both civil and military purposes, generic scientific, test, suborbital, dummy, supply, and even unknown payloads require their own modifiers (Rows 13–18, respectively). Note that the "unknown payload" modifier goes within the bottom section of the frame. To denote an "unknown orbit," the letters "UNK" rest at the top of the icon (Row 19)—in keeping with the convention of other orbital modifiers as shown in Table 4. Rows 20 and 21 account for satellites with more specific, dual scientific/military purposes, like geodetic and oceanography satellites.

As with scientific platforms, dual-use military and commercial communications systems also exist. Another existing capability that requires space symbology is direct communications (Row 22). Although "direct communications" does not have a doctrinal definition, as a symbol for space operations, the term may be interpreted as a satellite that communicates via cross-link with another satellite rather than relaying its signal to a ground station for further transmission. NASA's Tactical Data Relay Satellite System (TDRSS, voiced as "tea dress") provides cross-linked communications. The International Space Station, for example, may transmit to a TDRSS satellite, which will relay the message to another TDRSS satellite, which is in view of a ground station. In this way, the ISS is always able to communicate with Mission Control regardless of whether the station itself is on the opposite side of the world from a direct-downlink ground station. In the civilian sector, the Iridium constellation also employs cross-linking among its satellites, allowing global coverage for satellite phone users. The joint force made extensive use of Iridium phones throughout the Global War on Terror. For military purposes, depicting this capability in symbology is essential because such satellites are less dependent on direct downlink. Due to the global coverage of cross-linked relay satellite communications systems, these become prime targets to any adversary who wants to isolate opponents to just certain regions of the earth. As a result, employing them or countering them involves different operational considerations.

Finally, a modifier is necessary to account for satellites that exist primarily for the purpose of monitoring natural disasters. Most notably, the Disaster Monitoring Constellation provides imagery of natural disasters to enable more effective response. Because militaries often provide relief in the wake of earthquakes, hurricanes, tsunamis, tornados, or even nuclear accidents, it is important to know what civil "disaster" satellites are available to assist those efforts. Additionally, if active electronic warfare or orbital warfare activities are ongoing, forces will likely want to avoid interfering with disaster satellites, especially if they are needed for ongoing relief in another part of the theater.

Row	Description	Icon	Notes/Reference	Example Usage
1	Missile Warning Satellite	OPIR	Proposed	OPIR
2	Situational Awareness Satellite	SDA	Proposed	SDA
3	Reentry Vehicle	RV	Proposed	RV
4	Video Imagery Satellite		MIL-STD-2525D, Appendix D— Land Unit Symbols, Table D-IV, p. 175	
5	Debris		Proposed	MEO DEB

Row	Description	Icon	Notes/Reference	Example Usage
6	Refueling Satellite	K	MIL-STD-2525D, Appendix C—Air Symbols, Table C-III, p. 146	K
7	Missile Warning Satellite	OPIR	Proposed	OPIR
8	Situational Awareness Satellite	SDA	Proposed	SDA
9	Reentry Vehicle	RV	Proposed	RV
10	Video Imagery Satellite		MIL-STD-2525D, Appendix D—Land Unit Symbols, Table D-VI, p. 214	

Row	Description	Icon	Notes/Reference	Example Usage
11	Refueling Satellite		MIL-STD-2525D, Appendix C— Air Symbols, Table C-IV, p. 153	
12	Joint Military-Commercial Satellite		Proposed	
13	Scientific		Proposed	
14	Test Satellite		Proposed	
15	Suborbital Payload		Proposed	

Row	Description	Icon	Notes/Reference	Example Usage
16	Dummy Payload		Proposed	
17	Supply Spacecraft		Proposed	
18	Unknown Payload		Proposed	
19	Unknown Orbit		Proposed	
20	Geodetic Satellite		Proposed	

Row	Description	Icon	Notes/ Reference	Example Usage
21	Ocean-ography Satellite		Proposed	
22	Direct Com-munication		MIL-STD-2525D, Appendix C—Air Symbols, Table C-IV, p. 153	
23	Disaster		Proposed	

Table 7: Expanded Icons and Modifiers for On-Orbit Capabilities. Source: Table compiled by authors. The icon in Row 12 was originally published in Paul Szymanski, "Issues with the Integration of Space and Terrestrial Military Operations," *Wild Blue Yonder Online Journal* (June 2020). https://www.airuniversity.af.edu/Wild-Blue-Yonder/Article-Display/Article/2226268/issues-with-the-integration-of-space-and-terrestrial-military-operations/.

On-Orbit Means—Orbital Warfare

In early 2019, the Defense Intelligence Agency (DIA) released an unclassified report entitled *Challenges to Security in Space*. The diversity of possible orbital antisatellite capabilities suggested by the DIA may require additional modifiers to accompany the antisatellite weapon icon (shown in Table 4, Row 3). DIA assesses that such systems "could include payloads such as kinetic kill vehicles, radio-frequency jammers, lasers, chemical sprayers, high-power microwaves, and robotic mechanisms."* Of these potential capabilities, only the symbology to depict radio-frequency jammers exists. With creative application, one may adapt useful symbology from other domains, but some capabilities require altogether new symbology. As with most symbology, there is flexibility in representation of a unit or orbital warfare capability; the same system may be represented legitimately in multiple ways through different combinations of icons and modifiers. Table 8 shows these suggested icons and modifiers.

Radiofrequency Jammers (Table 8, Rows 1–3)

Several ways exist for depicting jammers. In air domain symbology, the "J" can be used as a modifier icon (Row 1) or as an icon within the domain frame (Row 2). These symbols are easily adapted for the space domain. MIL-STD2525D's Appendix J–Signals Intelligence provides for greater specificity in typing jammers—should the need arise—by adding an additional letter after the "J" in the modifier (for example, "JB" indicates a barrage jammer, and "JD" indicates a deceptive jammer).† Land domain symbology employs wavy lines (Table 8, Row 3) to depict the jamming function, or more generically, the icon "EW" on a spacecraft frame or as a modifier could depict an electronic warfare capability.‡ Row 3 shows the army's doctrinal jamming icon adapted for use with the space domain frame and the antisatellite icon. Row 4 shows the EW icon superimposed on an unknown spacecraft frame.

---

*   Defense Intelligence Agency, *Challenges to Security in Space* (Washington DC, 2019), 10, https:// www.dia.mil/Portals/27/Documents/News/Military%20Power%20Publications/Space_Th reat_ V14_020119_sm.pdf.

†   US DOD, MIL-STD-2525D (2014), 778.

‡   US Department of the Army, Army Doctrine Reference Publication (ADRP) 1-02, *Terms and Military Symbols* (Washington DC: Government Printing Office, 2011), 4–24.

High-Power Microwave and Laser Weapons

There is no existing symbology for a high-power microwave weapon in space although such a weapon could theoretically be a part of any domain's symbology. The antisatellite weapon icon with the modifier "HPM" is proposed (Table 8, Row 5).

Lasers

A symbol for lasers does exist as a land equipment icon.* This icon need only be superimposed on a spacecraft symbol to denote a satellite with a laser weapon (Row 6).

Alternatively, the three-letter modifier "LAS" may be used (Row 7). This modifier may be more appropriate for scientific or communications payloads that employ nonweaponized lasers that could potentially interfere with other space systems.

Chemical Sprayers

A satellite weapon with a chemical sprayer is perhaps one of the more exotic-sounding capabilities suggested in the DIA report. In effect, a satellite equipped with a chemical sprayer would approach a target satellite and spray its optical sensor, thereby rendering it useless. It is the orbital equivalent of using a can of spray paint to blacken the lens of a security camera. In addition, "painting" heat radiators, communications antennas, and solar panels would degrade their capabilities. This degradation can be temporary (if paints that degrade over time through sublimation are used) or permanent (for other sturdier paints and reactive acids). In such cases, a large amount of spray may not even be necessary. Painting just a small corner of a large solar panel could inhibit the satellite from collecting the necessary amount of solar energy or cause it to slowly start pinwheeling because of differential solar pressure.

---

*   US DOD, MIL-STD-2525D (2014), 265.

A modifier does exist in land symbology to denote a chemical weapon (the capital letter "C"). However, using that modifier may unjustly connote that the satellite is carrying a chemical weapon—a capability that not even the DIA has imagined. To avoid such associations, and to more closely approximate the function of spray paint, the modifier "PIT" is suggested (Row 8).

Robotic Mechanisms

Perhaps surprisingly, there is currently no icon or modifier in either MIL-STD-2525D or ADRP 1-02 to depict robotic grappling technology. A modifier such as "RBT" may suffice, but the robotic mechanism justifies its own icon for two reasons. First, a robotic mechanism may not always be an antisatellite system, so the "antisatellite weapon" icon is not always an appropriate one. However, the presence of a robotic mechanism on orbit demands that additional attention be drawn to the fact; use of a general "satellite" icon with a modifier may not draw sufficient attention. The proposed icon (Table 8, Row 9) takes the base "satellite" icon and adds an angled "arm" to the bottom of the satellite.

Kinetic Kill Vehicle (KKV)

The DIA report suggests that a kinetic kill vehicle could be housed as a payload on a host satellite and released in preparation for an attack on a target satellite.* Once released, the KKV becomes an independent satellite that performs the attack or strike function. Borrowing from air platform iconography, the "A" icon or the antisatellite weapon icon with an "A" modifier ("attack/strike") conveys the meaning of the KKV without inventing a new modifier (Rows 10 and 11, respectively). Alternatively, the letters "KKV" as a modifier provide a different option (Table 8, Row 12).

---

*   DIA, *Challenges to Security in Space* (2019), 10.

Mines

Although not mentioned in the DIA report, orbital mines are a type of space weapon contemplated since the Cold War. Mine warfare is a discipline in itself, and while it would be premature to create an entire lexicon of mine warfare for space operations (as the navy has for maritime operations), an inclusion of a mine symbol as a possible future threat is appropriate. Row 13 appropriates the ground symbol for an antitank mine, imposing it upon a space frame. Although there are other types of mines (moored, magnetic, antipersonnel, and so on), the black spot of the antitank mine is the simplest and suggests the intent to destroy equipment. Row 14 suggests the letters "MNE" as a modifier.

Row Number	Description	Icon/ Modifier	Notes/ Reference	Example Usage
1	Electronic Combat (EC) / Jammer [modifier]		MIL-STD-2525D, Appendix C—Air Symbols, Table C-IV, p. 154	
2	Jammer [icon]		MIL-STD-2525D, Appendix C—Air Symbols, Table C-III, p. 145; Appendix J— Signals Intelligence Symbols	
3	Jamming [modifier]		ADRP 1-02, Table 4-6, p. 4-25	

Row Number	Description	Icon/ Modifier	Notes/ Reference	Example Usage
4	Electronic Warfare [icon]		MIL-STD-2525D, Appendix D— Land Symbols, Table D-IV, p. 191	
5	High-Power Microwave [modifier]		Proposed	
6	Laser [icon]		MIL-STD-2525D, Appendix D— Land Symbols, Table D-XI, p. 265	
7	Laser [modifier]		Proposed	
8	Chemical Sprayer [modifier]		Proposed	

Row Number	Description	Icon/ Modifier	Notes/ Reference	Example Usage
9	Robotic Mechanism [icon]		Proposed	
	Robotic Mechanism (i.e., a "Grappler") [modifier]	GRP	Proposed	LEO GRP
10	Attack/ Strike [modifier]	A	MIL-STD-2525D, C—Air Symbols, Table C-IV, p. 152	A
11	Attack/ Strike [icon]	A	MIL-STD-2525D, C—Air Symbols, Table C-III, p. 145	A
12	Kinetic Kill Vehicle [modifier]	KKV	Proposed	KKV

Row Number	Description	Icon/ Modifier	Notes/ Reference	Example Usage
13	Mine [icon]		MIL-STD-2525D, Appendix D— Land Symbols, Table D-XI, p. 267	
14	Mine [modifier]	MNE	Proposed	GSO MNE

Table 8: Suggested Symbology for Potential Orbital Warfare Capabilities. Source: Table compiled by authors based on information in MIL-STD-2525D, Appendix D and ADRP 1-02, and authors' proposed new symbology for space.

Missiles

In their relation to space warfare, missiles require special consideration apart from other kinds of weapons. Not only is a basic knowledge of missile symbology essential for traditional missile warning support, but it is also necessary to consider existing antisatellite missile capabilities and the possibility of future space-based missile capabilities. Fortunately, the existing missile symbology is flexible enough to accommodate ground-to-space, space-to-space, or even space-to-ground missiles (Table 9).

The basic missile icon depicts a stylized, vertical missile (Row 1). Rather than placing modifiers above it or below it as with the satellite icons, however, missile modifiers go to the left or right side of the missile. In this system, a "B" placed on the right denotes that the missile has a ballistic trajectory (Row 2). The ballistic missile may be further defined by its range with modifiers listed in Rows 3–7: "SR" for short-range missiles (ranges of less than 1,000 km), "MR" for medium-range missiles (ranges between 1,000 and 3,500 km), "IR" for intermediate range missiles (also 1,000 to

3,500 km; a duplicate term for medium-range missiles), "LR" for long-range missiles (3,500 to 5,500 km), and "IC" for intercontinental missiles (ranges greater than 5,500 km). Not only are these modifiers useful within the missile warning function of space operations, but they would be useful to any headquarters engaged in space operations.

The modifiers described in the preceding paragraph are important but well-established. After all, ballistic missiles have been around for the better part of a century. With Row 8, we may begin to contemplate more recent and future capabilities. Row 8 shows the modifier for a missile whose destination is space, denoted by the letters "SP" on the right of the frame. A designated antisatellite missile launched from the ground would bear this marking. Transferring the symbol to the left-hand side (Row 9) denotes a missile whose launch origin is within space—a proposed capability at least since the Strategic Defense Initiative. Given these existing modifiers, it is interesting to note that military symbology allows for a space-to-space missile, even if policy or technology possibly do not (see Figure 22).

Row	Description	Icon/Modifier	Notes/Reference
1	Missile [icon]		MIL-STD-2525D, B—Space Symbols, Table B-VI, p. 138
2	Ballistic Class Missile [modifier]	**B**	MIL-STD-2525D, B—Space Symbols, Table BVII, p. 138

Row	Description	Icon/Modifier	Notes/Reference
3	Short-Range Missile [modifier]	S R	1,000 km or less; MIL-STD-2525D, B—Space Symbols, Table BVIII, p. 139
4	Medium-Range Missile [modifier]	M R	1,000 to 3,500 km; MIL-STD-2525D, B—Space Symbols, Table BVIII, p. 139
5	Intermedi-ate-Range Missile [modifier]	I R	1,000 to 3,500 km; MIL-STD-2525D, B—Space Symbols, Table BVIII, p. 139
6	Long-Range Missile [modifier]	L R	3,500 to 5,500 km; MIL-STD-2525D, B—Space Symbols, Table BVIII, p. 139
7	Intercontinental-Range Missile [modifier]	I C	5,500 km or more; MIL-STD-2525D, B—Space Symbols, Table BVIII, p. 139

Row	Description	Icon/Modifier	Notes/Reference
8	Space Launch Destination [modifier]		MIL-STD-2525D, B—Space Symbols, Table BVIII, p. 140
9	Space Launch Origin [modifier]		MIL-STD-2525D, B—Space Symbols, Table BVII, p. 138

Table 9: Useful Missile Symbology for Space Operations. Source: Table compiled by authors based on information from MIL-STD-2525D, Appendix B.

Figure 22: Symbol for a Missile with Origin and Destination in Space. *Source:* Created by Paul Szymanski based on existing icons and modifiers.

Space Operations—Ground Unit Symbology

Army symbology adopts the same conventions outlined in MIL-STD-2525D but offers some useful additions for space operations, particularly in consideration of how one might employ ground-based components in space operations. As shown in Table 1, blue rectangles denote friendly ground units, and red diamonds denote enemy ground units. An additional amplifier at the top of the rectangle denotes the echelon of the unit (see Table 10, Echelon Amplifiers). Although the air force did not routinely adopt army symbology for space operations in the past, space force units have an opportunity to employ a broader symbolic lexicon. A space force unit conducting space operations missions with a ground component would be represented by their equivalent army formation. For example, a space Delta commanded by a colonel would be analogous to an army brigade, whereas a subordinate squadron within the Delta would be more analogous to an army battalion.

AMPLIFIER	DESCRIPTION
Ø	TEAM/CREW
•	SQUAD
••	SECTION
•••	PLATOON/DETACHMENT
I	COMPANY/BATTERY/TROOP
I I	BATTALION/SQUADRON
I I I	REGIMENT/GROUP
X	BRIGADE
X X	DIVISION
X X X	CORPS
X X X X	ARMY
X X X X X	ARMY GROUP
X X X X X X	THEATER
+ +	COMMAND[1]

Notes: 1. Command is a unit or units, an organization, or an area under the command of one individual. It does not correspond to any of the other echelons.

Table 10: Echelon Amplifiers. Source: MIL-STD-2525D, p. 172.

The only symbol for a space unit currently used within the army is the rectangle with the "space" icon—a starburst (Row 1 of Table 11 below).* While useful as a generic symbol, much is lacking in the specificity of ground symbology. A second unit with a dedicated symbol is the "Signal—Tactical

---

*   This symbol was introduced in the 2016 version of ADRP 1-02—after the 2014 version of MIL-STD-2525D. As such, the symbol does not appear in MIL-STD-2525D.

Satellite" unit. While the assigned nomenclature is awkward, the icon lends itself to units such as the US Army Satellite Operations Brigade and its subordinate elements. As for other space mission units, there are no dedicated land unit symbols, but this problem can be easily addressed by combining space domain icons with the land domain unit frame. In simplest terms, every orbital mission listed here has a ground component either to command and control the orbital segment of the space system or to process the data that the space segment provides. Furthermore, offensive and defensive space electronic warfare units require unique symbology.

By combining the satellite type icons of Table 4 with the ground unit rectangle, we may arrive at intuitive unit symbology. Nothing precludes such combination; in fact, within the system of military symbology, the icons—while primarily useful in one domain—are domain agnostic. The frame specifies the domain-specific context in which the icon is being used. The innovation here is not in the creation of new symbology but in combining existing elements in ways that are not typically done. For example, a unit that controls GPS satellites can be depicted by combining the "Navigation Satellite" icon and the friendly ground unit frame (Row 2). By such combinations, we may depict all types of ground segment units.

The space force's offensive space electronic warfare units—here defined as those units that seek to control the link segment of space systems through offensive action—again require creative combination.* For such ground-based systems, the "Electronic Combat (EC)/Jammer" modifier from the air domain, or the "Jamming" function icon from the land domain could feasibly combine with the space ("star") icon (Rows 4-5). The jamming function icon (the "wavy lines") of Row 5 seems superior to the "J" modifier because the jamming function icon visually suggests the activity involved. These jammers could be further modified by any of the doctrinal modifiers for jammers or with additional modifiers or amplifiers that specify the capabilities of the unit's equipment.

*   "US Space Force locks on to 'Enemy' Satellite," *Times*, April 20, 2020, https://www.thetimes.co.uk/edition/world/us-space-force-locks-on-to-enemy-satellitesp88h0pl3q, accessed April 21, 2020.

Similarly, defensive space control units—here defined as those units that seek to control the link segment of space systems through defensive action—do not lend themselves to a ready solution. Army military intelligence symbology provides electronic warfare symbols for direction finding, electronic ranging, intercept, and search, but none of these quite hit the mark when considering attempts of space operations units to monitor, detect, characterize, and geolocate sources of electromagnetic interference and other antisatellite systems.* There are symbols for "control" and "sensor," but the first implies a relationship with movement control while the second is too nonspecific. Perhaps the simplest solution is to apply the modifier "DSC" to the space icon (Row 6).

Row 7 proposes a symbol for a spaceport, such as Spaceport America in New Mexico. This symbol combines the transportation wheel with the space star, which is in keeping with the way other domain forces depict their ports. The icon for an airport, for example, combines the transportation wheel with a stylized runway atop it. Similarly, the icon for seaport combines the transportation wheel and an anchor. As an installation, the complete spaceport symbol includes the black rectangle amplifier at the top.

Row Number	Description	Icon/ Modifier	Notes/ Reference	Example Usage
1	Space Unit		ADRP 1-02, Table 4-1, p. 4–5	

---

*   See Table 4-6, p. 4–25 of ADRP 1-02.

Row Number	Description	Icon/ Modifier	Notes/ Reference	Example Usage
2	Signal— Tactical Satellite		MIL-STD-2525D, Table DIV, p. 175	
3	Navigation Satellite Unit		Suggested combination based on MIL-STD-2525D, Appendix B	
4	Offensive Space Electronic Warfare Unit		Suggested combination based on MIL-STD-2525D, Appendix C, and ADRP 1-02	
5	Offensive Space Electronic Warfare Unit		Suggested combination based on MIL-STD-2525D, Appendix C— Land Unit Symbols, Table D-IV, p. 192	
6	Defensive Space Control Unit		Proposed	

Row Number	Description	Icon/ Modifier	Notes/ Reference	Example Usage
7	Spaceport		Suggested combination based on MIL-STD-2525D, Appendix C— Land Unit Symbols, Table D-XIII, p. 289	

Table 11: Some Useful Symbology for Ground-Based Space Units from the Army's Lexicon. Source: Table compiled by authors based on information from MIL-STD-2525D, Appendices C and D; from ADRP 1-02; and from the authors' proposed new symbology for space.

Ground–Based Equipment Symbology

Along with the depiction of ground unit symbology, the system also allows for the depiction of individual pieces of equipment. After all, as with satellites themselves, there are some pieces of ground-based equipment that the space operator may desire to track or target. The pieces of equipment in Table 12 fall into the sensor category—optical, infrared, signals intelligence, or radar sensors. However, many of the unitized icons and modifiers shown in previous tables could be paired with the ground equipment frame (the blue circle) to convey meaning.

Row 1 depicts a friendly, ground-based sensor. The small circles under the sensor platform suggest wheels and indicate that the sensor is mobile. In this case, the letters "SP" are used as a modifier to indicate that the sensor is space-related—an alternative to using the star modifier. The "O" at the bottom denotes an optical sensor. In a similar manner, Rows 2–4 provide examples of infrared (IR), signals intelligence (SI), and radar platforms.

Row	Symbol	Meaning
1		Friendly, Ground-Based Sensor—Mobile, Space-Related, Optical
2		Enemy, Ground-Based Sensor—Mobile, Space-Related, Infrared
3		Neutral, Ground-Based Sensor—Mobile, Space-Related, Signals Intelligence
4		Unknown Affiliation, Ground-Based Sensor—Mobile, Space-Related, Radar

Table 12: Some Ground-Based Sensor Equipment Symbols. Source: Created by authors.

Symbology for Installations

In addition to symbols for ground-based equipment, some basic symbology is required to depict installation activities necessary for the conduct of space warfare. These symbols employ land installation frames and combine established icons (see MIL-STD-2525D Table D-XIII). As in the previous section, the "SP" modifier denotes space-related activities. Row 1 uses the modifier "FUE" to denote a generic fuel production installation; Row 2 the letters "AST" to denote antisatellite (ASAT) manufacturing facilities; Row 3 the letters "MFC" to denote generic manufacturing related to space systems, such as satellite builders; Row 4 the letters "RAD" to stand for "radiation"; and Row 5 the letters "TOX" for toxic fuels. Unless the intent is to unleash radioactive or toxic substances into the environment, "No Fire Areas" will likely surround these facilities.

Row	Symbol	Meaning
1	SP FUE	Enemy Installation, Space-Related Hazardous Material Production, Rocket Fuel
2	SP AST	Unknown Affiliation, Space-Related Weapons Manufacture
3	SP MFC	Friendly Installation, Missile and Space Production Facility
4	SP RAD	Neutral Installation, Space-Related Radioactive Storage

Row	Symbol	Meaning
5	SP / REL	Neutral Installation, Space-Related Toxic Release Inventory
6	SP / AIR	Neutral Installation, Space-Related Airport
7	SP / SPP	Neutral Installation, Specific Spaceport

Table 13: Space-Related Installation Symbology. Source: Created by authors from existing icons and modifiers.

Rules of Engagement Amplifiers

In future conflicts that may involve a myriad of orbital and terrestrial weapons systems, it will be important for space operators to know what rules of engagement (ROE) are authorized against a particular system. In space warfare, activities may escalate from mere deterrence to destruction and are defined in Table 14.

Term	Definition
Deter	Deterrence operations convince adversaries not to take actions that threaten US vital interests by means of decisive influence over their decision-making. Decisive influence is achieved by credibly threatening to deny benefits and/or impose costs while encouraging restraint by convincing the actor that restraint will result in an acceptable outcome.
Deception	Those measures designed to mislead the adversary by manipulation, distortion, or falsification of evidence to induce the adversary to react in a manner prejudicial to their interests.
Disruption	The temporary impairment of the utility of space systems, usually without physical damage to the space segments. These operations include delaying critical mission data support to an adversary. Given the perishability of information required to effectively command and control military operations, this disruption impedes the effective application or exploitation of that data. Examples of this type of operation include jamming or refusing or withholding data support or spare parts.
Denial	The temporary elimination of the utility of the space systems, usually without physical damage. This objective is accomplished by such measures as denying electrical power to the space ground nodes or computer centers where data and information are processed and stored.
Degradation	The permanent impairment of the utility of space systems, usually with physical damage. This option may include attacks against the terrestrial or space element of the space system. For example, a ground-based laser could be used to damage the optics of an imaging sensor without impairing other functions of the satellite bus.
Destruction	The permanent elimination of the utility of space systems, usually with physical damage. This last option includes special operations forces missions to interdict critical ground nodes, bombing uplink/downlink facilities, and attacks against space elements with either kinetic-kill or directed-energy weapons.

Table 14: Definitions: Space Defense Definitions. Source: Definition for "deterrence" from US Department of Defense, *Deterrence Operations Joint Operating Concept* (Washington, DC, 2006), 3, https://www.jcs.mil/Portals/36/Documents/Doctrine/concepts/joc_deterrence.pdf?ver=2017-1228-162015-337. Other definitions derived from terms listed in Air Force Doctrine Document 2-2, *Space Operations* (Washington, DC, 2006), 23.

Although the two-letter amplifiers discussed in Figure 15 provide a textual way of depicting ROE, Table 15 depicts this escalation with a series of visual amplifiers that sit atop the base icon—in this case, a neutral reconnaissance satellite. Actions that allow deterrence require no amplifier (Row 1). In fact, this may be considered a default rule of engagement. A series of varying shapes with differing colors amplify the icon to allow deception, disruption, denial, degradation, and destruction (Rows 2 through 6, respectively). Importantly, although Table 15 shows these amplifiers paired with space frames, they might also be paired with event frames to signify a change in ROE type (see section on "Event Symbols").

Row	Sample Icon with Amplifier	ROE Type
1		Actions that DETER only
2		Attacks allowed up to DECEPTION
3		Attacks allowed up to DISRUPTION

Row	Sample Icon with Amplifier	ROE Type
4		Attacks allowed up to DENIAL
5		Attacks allowed up to DEGRADATION
6		Attacks allowed up to DESTRUCTION

Table 15: Proposed Amplifiers to Signify Authorized Rules of Engagement against Satellites. Source: Created by authors.

Section 3: Command and Control Lines, Areas, and Points

In the space domain, the implementation of control measures—"means of regulating forces or warfighting functions"—is not fundamentally different than in any other domains.* In any domain, such regulation involves two complementary efforts: first, a combination of anticipating and knowing the locations of forces and second, planning for the interactions of forces in relation to other forces—friendly, enemy, and neutral—within the operating

---

* US Department of the Army, Army Doctrine Reference Publication (ADRP) 1-02, *Terms and Military Symbols* (Washington DC: Government Printing Office, 2011), 1–22.

environment. To do this effectively, the plan must divide the operating area into manageable sections that allow for the command and control (C2) of forces within them. Command and control symbols—lines, areas, and points—form the basis of this lexicon.

Despite the commonalities that all domains share, however, applying the existing symbolic language to space operations requires a significant cognitive leap for three important reasons. First, as an area of responsibility, space is enormous. Between the von Karman line at an altitude of 100 kilometers and geosynchronous (GEO) orbit at 35,786 kilometers reside roughly 200 quadrillion (200,000 trillion) cubic kilometers of space—a volume about 140,000 times as voluminous as all of the earth's oceans. Second, on orbit, satellites adhere to the rules of orbital motion, which are not as intuitive as the rules of motion on land or in the air. Within this vast volume of space, satellites are constantly moving (even at geosynchronous altitudes) so "fixed" volumes of outer space are difficult to visualize and do not follow perceptions learned from living on the earth. Because of the unique properties of orbital motion, an antisatellite on the opposite side of the earth in a similar altitude and inclination orbit is actually "close" (as defined as the amount of time and fuel it requires) to reach its intended target. As a historical note, these counterintuitive properties of orbital motion manifested themselves during the Apollo program when some of the astronauts had trouble completing manual rendezvous. When they "stepped on the gas" to get closer, they actually made themselves maneuver into a higher orbit than their intended target.

Orbits and their characteristics have implications for their usefulness to forces in other domains, in many cases providing windows of access rather than continuous access. As a consequence, operating areas become dynamic, aligned to a specific theater, a specific mission, or perhaps both. Third, although satellites operate on orbit, space operations are inherently multidomain. They involve ground segment operations, use of the electromagnetic spectrum, and the transit of objects and people into and out of the atmosphere. Therefore, all space operations require simultaneous C2 in three operating areas—tellurian (meaning all earthbound domains), electromagnetic, and orbital. In the future, operating areas such as the moon and Mars may be considered. Each may be subdivided depending on the needs of the operation, but they remain like links of a chain, the weakest inducing the greatest operational vulnerability.

Fortunately, for a given operation, not all of space, the earth's surface, or the electromagnetic spectrum are of equal importance at all times. Making such areas operationally useful requires implementation of command-and-control lines and areas—a subset of control measures that govern the division of larger operating areas such as areas of responsibility (AORs), joint operations areas (JOAs), or areas of operations (AOs).* As in ground, air, or maritime operations, space operations require synchronized activity across multiple operating areas. By applying boundaries, areas of operations, named areas of interest (NAIs), and target areas of interest (TAIs) to the peculiarities of the space domain, we arrive at a more effective way of conceptualizing space operations not only for the space practitioner but for the joint force as a whole. In so doing, however, one should also be aware of the potential vulnerabilities created. As in ground military operations, many times the most vulnerable points to attack are at the boundaries between AORs, JOAs, and AOs, because these kinds of attacks can more readily create confusion in command relationships and thus impede effective responses.

Command and Control Lines and Areas

The category of C2 graphics useful for space operations—both in space and from the surface—requires first considering lines and areas as tools for defining the utility of space. Table 16 lists the most fundamental of these. While the military standard for joint symbology provides additional detail, in general, lines and areas may belong to friendly or enemy forces; they may be in place, planned, or on order; and they may have names or other amplifying information associated with them. Boundary lines may be lateral, forward, or rearward boundaries. As in other domains, lines

---

*　In the geometric sense, of course, areas—even for land operations—are properly "volumes of interest," but since three-dimensionality is taken to be a given in space, the traditional terminology is sufficient to overcome the authors' pedantry.

and areas provide the means by which to establish operating areas and to regulate operations within those areas.

For orbital operations, the discussion of boundaries begins with the designated areas of responsibility (AORs) that the president assigns to the geographic combatant commanders. Figure 23 shows a partial map of global AOR boundaries. The commander of US European Command (EUCOM), for example, is responsible for an AOR that stretches from the mid-Atlantic in the west to the edge of Russian waters in the east. In the south, EUCOM borders US Africa Command (green in the graphic below), US Central Command (CENTCOM), and US Indo-Pacific Command (blue in the graphic below). In the north, it shares responsibilities for the Arctic with US Northern Command. Per the 2019 Unified Command Plan, the commander of US Space Command has responsibility for the volume of space beginning 100 kilometers above the surface of the earth. Since each terrestrial combatant commander—the commander of European Command, for example—bears responsibility for both the ground and the airspace (up to 100 kilometers) across the breadth of his AOR, an understanding of boundary designation is essential.

Consider, for example, a satellite reentering earth's atmosphere. The object remains within Space Command's AOR while above 100 km. Globally dispersed, ground-based tracking radars and C2 elements leverage various portions of the electromagnetic spectrum to gather information about the satellite and communicate its whereabouts. When the satellite leaves orbital space, it may enter the atmosphere above Africa, in Africa Command's AOR. As the satellite loses altitude, it continues to travel eastward, ultimately landing in the South Pacific, the AOR of Indo-Pacific Command. Africa Command would certainly be interested in the deorbit because of the potential danger from falling debris or vaporized hazardous satellite fuel. Indo-Pacific Command would be interested for the same reasons, and even though the satellite would likely land in the water, it would take actions necessary to clear maritime or air traffic from the anticipated impact site.

Less routine activities further illustrate the importance of well-understood control measures. First, consider a missile that launches from the ground and travels above the von Karmen line. This could be an antisatellite missile, the first stage of a hypersonic glide vehicle, or a nuclear-tipped missile set to burst in space. In these cases, multiple commanders may bear responsibility for prelaunch and postlaunch military action; air strikes, special operations activity, cyberattack, or boost-phase intercept actions could feasibly fall under four separate combatant commands.

Space Command would bear the responsibility for providing space-based warning of the launch, for protecting on-orbit capabilities, and for coordinating with affected commands during and after the attack. The special case of a hypersonic glide vehicle potentially combines the difficulties of a satellite reentry and a maneuverable missile attack.

Figure 23: A Partial Map of AOR Boundaries among Geographic Combatant Commands. *Source:* Adapted image from US Department of Defense, "Unified Combatant Commands," https:// upload.wikimedia.org/wikipedia/commons/e/e6/Unified_Combatant_Commands_map.png.*

On traditional, flat maps, the boundary graphic is represented by a simple line that connects with other boundary lines. An amplifier denotes the echelon. In Row 1, the echelon amplifier is "XXXXXX," denoting an AOR boundary, that is, a boundary between combatant commands. An anticipated boundary—one that is not yet established but is planned or will be executed on order—is represented by a dashed line (Row 2). The echelon amplifier in Row 2 is a single "X," denoting a brigade or Delta

---

*   US Department of Defense, "Unified Combatant Commands," accessed September 20, 2020, https://upload.wikimedia.org/wikipedia/commons/e/e6/Unified_Combatant_Commands_map. png.

boundary. In the three-dimensional applications of space warfare, these simple boundary graphics may look less like lines and more like planes— vertical walls between terrestrial AORs with arched ceilings between each terrestrial AOR and the space AOR.

Because AORs are so large, it is often useful to establish smaller, internal operating areas. To do this, the basic C2 symbol that is needed is the area of operations (AO). The AO can take on whatever shape is deemed necessary to meet the requirements of the operation; it may connect to other unit AOs, be part of a higher headquarters AO, or stand alone noncontiguously. Row 3 of Table 16 shows an AO named "AO Buffalo." The following section, "Divvying Up the Space AOR," goes into greater detail on how one might implement such divisions using different varieties of AOs.

A third type of C2 area is the named area of interest (NAI)—a "geographical area where information is gathered to satisfy specific intelligence requirements" (Row 3).* Generically, an NAI is something like an operating area that commanders establish to govern intelligence collection activity in support of their information requirements. Within the realm of space operations, an NAI may be tied to an object (such as a satellite), or it may be tied to a physical location on ground, on orbit, or even within the electromagnetic spectrum. A target area of interest (TAI) is similar to an NAI, but it has weapons allocated against it to engage enemy targets within the area (Row 4). An NAI may transition into a TAI and vice versa. For the ground segment of the space domain, the preceding command and control areas are sufficient as is. Within the orbital and electromagnetic segments of the space domain, however, the concepts require adaptation.

---

*   US DOD, MIL-STD-2525D (2014), 400.

Row	Description	Notes/Reference	Example Usage
1	Boundary	MIL-STD-2525D, Appendix H—Control Measure Symbols, Table H-III, p. 395	SPACECOM — XXXXXX — EUCOM
2	Planned or On-Order Boundary	MIL-STD-2525D, Appendix H—Control Measure Symbols, Table H-III, p. 395	------------ x ------------
3	Area of Operations	MIL-STD-2525D, Appendix H—Control Measure Symbols, Table H-V, p. 399	AO Buffalo
4	Named Area of Interest	MIL-STD-2525D, Appendix H—Control Measure Symbols, Table H-V, p. 400	NAI 1
5	Target Area of Interest	MIL-STD-2525D, Appendix H—Control Measure Symbols, Table H-V, p. 400	TAI 1

Table 16: Command and Control Areas. Source: Table compiled by authors from information in MIL-STD-2525D, Appendix H.

Dividing the space AOR into smaller AOs for subordinate commanders to manage will be necessary for effective command and control. Within each AO, additional lines and areas may create boundaries within boundaries to focus the efforts of the limited resources available to the AO commander.* Rather than referring to the region as a generic AO, we may specifically designate this region as a space defense region (SDR) with subordinate areas, divisions, or sectors. Within SDR GEO-EUR, for example, it may also be appropriate to designate NAIs around friendly assets, around a suspicious enemy asset, or even around empty orbital slots into which an adversary is anticipated to place an asset. The establishment of these NAIs will drive prioritization of reconnaissance assets. If a friendly counterspace asset is placed against the enemy asset, it becomes a TAI.

Since NAIs and TAIs are concepts from land-domain operations, their applicability to the ground segment of space operations should not come as a surprise. Indeed, all commands should consider NAIs and TAIs for enemy surface assets that affect the space domain. NAIs may cover enemy launch or ground control facilities, satellite and rocket fuel manufacturing plants, known space force garrisons, anticipated operating areas of deployable counterspace forces, or even over the footprints of SATCOM beams (this last use being a proxy for control measures within the EM spectrum). Depending on the circumstance, weapons could be allocated against any of the preceding.

The third component of a space system—the link segment—requires NAIs and TAIs also, but the visualization techniques are necessarily different because the EM spectrum is not a geographic region. The EM spectrum does, however, exist in physical space, so with the proper approximation (i.e., the proper kind of map), the graphics can be applied to it. A technique for designating NAIs or TAIs involves overlaying the NAI and TAI graphics on top of a visualization of the electromagnetic spectrum.

---

* In the geometric sense, of course, areas of interest in space are properly "volumes of interest," but since three-dimensionality is taken to be a given, we keep with the traditional terminology rather than arguing for space-specific terminology.

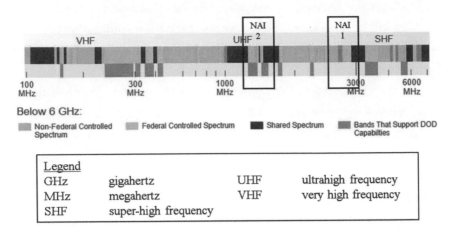

Figure 24: Electromagnetic Spectrum with Designated NAIs. *Source:* Created by Jerry Drew with base image of spectrum adapted from Joint Publication 6-01, *Joint Electromagnetic Spectrum Management Operations*, (Washington, DC, 2012), I-3.

Figure 24 shows a portion of the electromagnetic spectrum that may be of military interest. In this figure, NAI 1 covers a portion of the spectrum commonly used to downlink data from Department of Defense satellites (between 2200 and 2290 MHz).* Depending on the criticality of this portion of the spectrum, a commander may dedicate assets to observe this portion of the spectrum for enemy activity. For the enemy, of course, this portion of the spectrum might be a TAI—a potential target for offensive action. Similar to NAI 1, NAI 2 is valuable because the spectrum from 1219–1390 MHz contains a portion of the spectrum used by GPS. This spectral NAI, then, has a direct tie to the portion of orbital space designated as AO GPS.

While it is certainly true that, depending on the required capability in need of protection, enemy capabilities arrayed against it, and the time that the NAI is assessed to be vulnerable, the properties of the NAI would be further refined. Still, the illustration demonstrates the utility of NAIs and TAIs in concept and the need for consideration of the EM spectrum as one would consider ground-based or orbital AOs. Indeed, the EM spectrum is not only an aspect of the AO; it is an AO in itself, and treating it as such ensures that it is a deliberate part of planning and execution. In

---

*   US Joint Staff, JP 6-01, *Joint Electromagnetic Spectrum Management Operations 2012*, I-3.

space operations, one cannot take for granted the spectrum's availability, and without it, the ground segment and the space segment are isolated from each other.

Divvying Up the Space AOR

As discussed earlier, the vastness of the space AOR and the need for coordination lends itself to a consideration of how the entirety of the AOR may be broken down into manageable segments—just as a terrestrial AOR would be broken down into various kinds of operating areas. Symbols to depict such divisions are only part of the problem; the second part of the problem is deciding upon an organizing principle within which to employ the symbols. In general, the space-internal boundaries could follow one of three organizing principles. In the first line of logic, terrestrial AOR boundaries provide the template for in-space boundaries. In a second, boundaries surround functional missions. In a third, boundaries center on orbital altitudes. A fourth type of AO is the least intuitive and requires special graphical representation. These AOs are defined by altitudes and inclinations, the two most critical aspects of satellite orbits, and are represented in a special graph called an "Altitude-Inclination Survey." For an operational-level application of space warfare, some combination of the four organizing principles will be necessary.

In the first conception, the edges of the tellurian AOR extrude vertically to the 100-km line, and the projection of these boundaries continues through space out to geosynchronous orbit (or beyond). In this case, European Command would be responsible for its AOR as always, and a separate commander for space operations could be responsible for the orbital operating area above Europe. Figure 25 shows the projection of Earth-surface AOR boundaries onto the globe of geosynchronous orbit. This construct may be useful if the EUCOM commander requires protection from an adversary's satellites or desires to negate an adversary's capabilities as they transit over his AOR.

The shortcoming to this organizing principle is that the geosynchronous satellites most useful to the EUCOM commander (or to the enemy) are not directly above the EUCOM AOR—in the AO we may call "AO EUR-Space."

They are, of course, roughly over the equator, at the tip of the white arrow in Figure 25. It would therefore be necessary to maintain close coordination with adjacent AO commanders—perhaps AO AFRICA-Space and AO CENT-Space—as well as with Space Command itself. If not assigned directly to European Command, the commander of AO EUR-Space would certainly enjoy a close relationship with EUCOM, possibly even coming under that commander's control in certain scenarios.

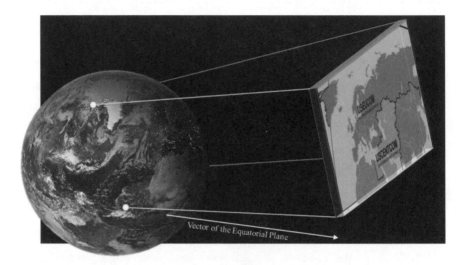

Figure 25: A Projection of Terrestrial AOR Boundaries into Space. *Source:* Created by Jerry Drew with base Earth image from NASA, https://www.nasa.gov/50th/50th_magazine/earthSciences.html "Earth Science: NASA's Mission to Our Home Planet, https://www.nasa.gov/50th/50th_magazine/earthSciences.html and modified AOR map from Department of Defense, "Unified Combatant Commands," https://upload.wikimedia.org/wikipedia/commons/e/e6/Unified_Combatant_Commands_map.png.

For nongeosynchronous satellites, the problem is more nuanced, and a separate conception of operating areas is necessary—one based on function. In this conception, a subordinate commander is responsible for an AO defined by a lower and an upper altitude—a spherical shell perhaps several hundred or even several thousand kilometers thick. Think, for example, about the Global Positioning System (GPS) constellation. With a minimum of twenty-four satellites operating at altitudes of roughly 20,200 km, the subordinate commander responsible for that constellation could have a global AO defined around the altitude parameter. Since GPS satellites transit across the earth's surface, different satellites are valuable to different combatant commands at different times. The commander of "AO-GPS" may remain in general support to most combatant commands most of the time with periods of direct support for certain tasks (employing flex power in support of an operation, for example). And although AO-GPS transits through AO EUR-Space, AO AFRICA-Space, and AO CENT-Space, AO-GPS would not be a part of any of those AOs. There would, however, be a need for well-established supporting relationships, control measures, and extensive cooperation among elements. Figure 26 depicts a cross section of AO GPS in relation to those AOs aligned with surface AORs.

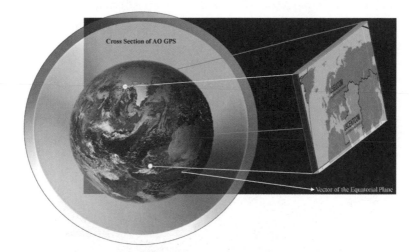

Figure 26: AO GPS in Relation to the Projection of Terrestrial AOR Boundaries.
Source: Created by Jerry Drew with base Earth image from NASA, https://www.nasa.gov/50th/50th_magazine/earthSciences.html "Earth Science: NASA's Mission to Our Home Planet," https://www.nasa.gov/50th/50th_magazine/earthSciences.html and modified AOR map from Department of Defense, "Unified Combatant Commands," https://upload.wikimedia.org/wikipedia/commons/e/e6/Unified_Combatant_Commands_map.png.

A third conception combines elements of the first two but relies more on orbital location than function. Consider the geosynchronous portion of the space AOR as a ring-shaped AOR bounded several degrees to the north and south of the equator with an average altitude of about 35,786 km. The commander of "AO GEO" would certainly have her hands full trying to wrangle the entirety of that ring and its varied mission sets. In fact, under the current organization concept, such an attempt would probably not be possible for traditional space support missions because commands are aligned to functions (like communications or missile warning, for example). For the orbital warfare mission, however, an AO GEO might be very practical, if the ring was sliced into manageable segments for subordinate commanders to manage. AO-GEO could be divided into AO GEO-AFRICA, AO GEO-PAC, or even an AO for the eastern hemisphere. Each segment of the larger AO GEO would be in shape like the crust of a pizza slice and, depending upon the threat picture, may be further designated as a space defense identification zone (more on these shortly). Figure 27 shows AO GEO-Eastern as a standalone operating area that combines portions of the GEO belt over Europe, Africa, and Asia. Such an AO may be permanently established, or it may become established in the event of a contingency (a planned or on-order boundary).

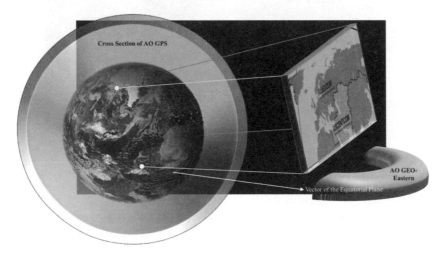

Figure 27: AO GEO-Eastern in Relation to AO GPS and the Projection of Terrestrial AOR Boundaries. *Source:* Created by Jerry Drew with base Earth image from NASA, "Earth Science: NASA's Mission to Our Home Planet," https://www.nasa.gov/50th/50th_magazine/earthSciences. html modified AOR map from Department of Defense, "Unified Combatant Commands," https:// upload.wikimedia.org/wikipedia/commons/e/e6/Unified_Combatant_Commands_map.png.

On a two-dimensional map, divvying up the geosynchronous belt into areas of operation may look something like the following depiction (Figure 28). This graphic divides the entirety of geosynchronous orbit into four space defense regions (SDRs), a term based on the historical lexicon of air defense. These regions are, from left (west) to right (east) as follows: SDR-US, SDR-Europe, SDR-Middle East, and SDR-Asia. The graphic depicts the most relevant satellites to global military operations (for the scenario for which it was created) and designates friendly and allied satellites in blue, potential adversary satellites in red, neutral satellites used by the United States in yellow, and neutral satellites in green. Importantly, the concept of the SDR can be used with any method of AO creation. However, once SDRs (and the corresponding command relationships within them) are established, adjustments will be difficult. As with any movements at GEO, changing satellite locations is costly in terms of time and on-board fuel.

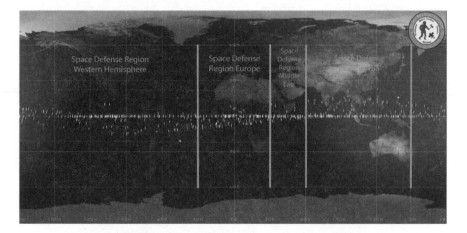

Figure 28: Space Defense Regions as Areas of Operation. *Source:* Graphic created by Paul Szymanski using the Satellite Orbit Analysis Program (SOAP) from The Aerospace Corporation and orbital data from https://www.space-track.org.

Region	Region Definition
SDR GEO	Space Defense Region Geosynchronous
SDR GEO ASIA	Space Defense Region Geosynchronous over Asia
SDR GEO EU	Space Defense Region Geosynchronous over Europe
SDR GEO ME	Space Defense Region Geosynchronous over the Middle East
SDR GEO US	Space Defense Region Geosynchronous over the United States

Region	Region Definition
SDR GEO-G-A	Space Defense Region Graveyard Orbit above Geosynchronous
SDR GEO-G-B	Space Defense Region Graveyard Orbit below Geosynchronous
SDR GEO-I	Space Defense Region Geosynchronous Inclined
SDR HEO	Space Defense Region above Geosynchronous (High Earth Orbit)
SDR LEO-E	Space Defense Region Low Earth Orbit Highly Eccentric
SDR LEO-H	Space Defense Region Low Earth Orbit—High (>600 and <5,000 km)
SDR LEO-L	Space Defense Region Low Earth Orbit—Low (<=500 km)
SDR LEO-M	Space Defense Region Low Earth Orbit—Medium (>500 and <=600 km)
SDR LEO-R	Space Defense Region Low Earth Orbit Retrograde
SDR LEO-S	Space Defense Region Low Earth Orbit Sun-Synchronous
SDR MEO	Space Defense Region Medium Earth Orbit (>=5,000 and <25,000 km)
SDR MOLY	Space Defense Region Molniya
SDR NOE	Space Defense Region No Orbital Elements

Table 17: Hypothetical Space Defense Regions and Their Definitions. Source: Created by Paul Szymanski.

The final method for divvying up an AOR involves mapping the positions of satellites by plotting their altitudes and inclinations. Figure 29 shows such a survey. Groups of satellites with similar altitudes and inclinations form user-defined space defense identification zones (SDIZs). "SDIZ-Blue Zone-LEO," for example, designates an area with multiple, high-value, friendly satellites operating in low-earth orbit (bottom left of the figure). While the SDIZs are defined in relation to orbital parameters, additional engagement zones may be defined in relation to a particular satellite that the enemy may target.

Within a greater space defense identification zone (SDIZ), the volumes of space around a target satellite may be further subdivided into long-range engagement zones (LREZ)—the larger black box, which indicates that the threat is greater than 10 kilometers or 20 minutes away. Similarly, the smaller black box defines the close-attack engagement zone (CAEZ). The CAEZ may be further subdivided into yellow and red categories. In the CAEZ-Yellow, the attacking spacecraft is 10 kilometers or less or 20 minutes or less from its target. In a red CAEZ, the attacker is within one kilometer or five minutes of its target. We will explore the utility of these divisions throughout the rest of the text.

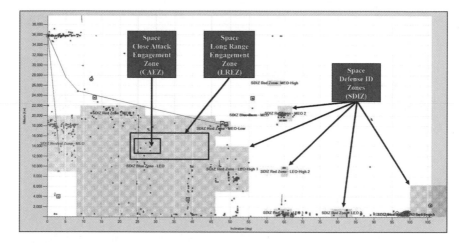

Figure 29: An Altitude-Inclination Survey with Space Defense Identification Zones. *Source:* Paul Szymanski, "Issues with the Integration of Space and Terrestrial Military Operations," *Wild Blue Yonder Online Journal* (June 2020). https://www.airuniversity.af.edu/Wild-Blue-Yonder/Article-Display/Article/2226268/issues-with-the-integration-of-space-and-terrestrial-military-operations/.

Importantly, SDIZs are the largest subdivision in the altitude-elevation plot and are conceptually linked to strategic courses of action, which may contain multiple SDIZs (Figure 30.). Similarly, LREZ's have a relationship with the operational level, supporting the possibility of many tactical courses of action. Each LREZ may contain multiple CAEZs. Within the CAEZs, tactical activities occur. In each zone, activities are deconflicted with other courses of action in adjacent and higher-echelon plans. At the strategic level, activities within the SDIZ must be deconflicted with terrestrial planning efforts, but that is not to say that such deconfliction cannot happen among tactical space and surface units. In fact, if a ground-based EW unit is in support of the orbital asset under attack, such coordination would be essential.

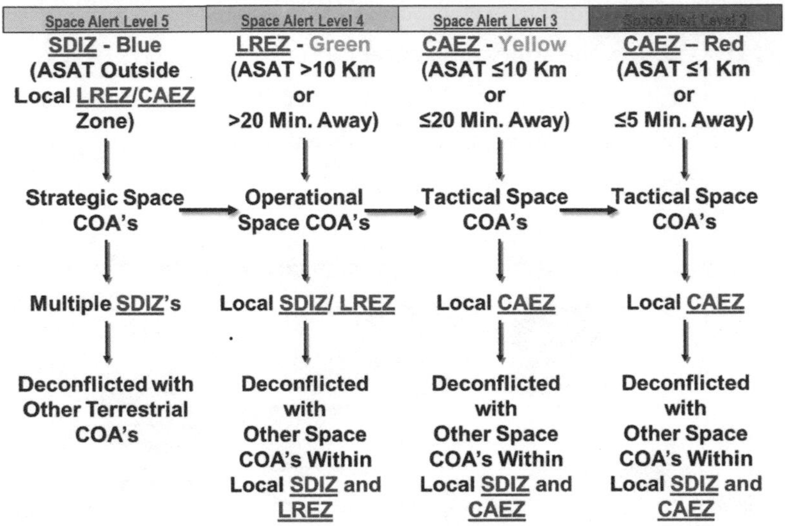

Space Alert Level 5	Space Alert Level 4	Space Alert Level 3	Space Alert Level 2
SDIZ - Blue (ASAT Outside Local LREZ/CAEZ Zone)	**LREZ - Green** (ASAT >10 Km or >20 Min. Away)	**CAEZ - Yellow** (ASAT ≤10 Km or ≤20 Min. Away)	**CAEZ – Red** (ASAT ≤1 Km or ≤5 Min. Away)
↓	↓	↓	↓
Strategic Space COA's →	**Operational Space COA's** →	**Tactical Space COA's** →	**Tactical Space COA's**
↓	↓	↓	↓
Multiple SDIZ's	**Local SDIZ/ LREZ**	**Local CAEZ**	**Local CAEZ**
↓	↓	↓	↓
Deconflicted with Other Terrestrial COA's	**Deconflicted with Other Space COA's Within Local SDIZ and LREZ**	**Deconflicted with Other Space COA's Within Local SDIZ and CAEZ**	**Deconflicted with Other Space COA's Within Local SDIZ and CAEZ**

Figure 30: Relationships of SDIZs, LREZs, and CAEZs across the Levels of War. *Source:* Created by authors.

Points

The third type of command-and-control graphics are point graphics (Table 18). Like boundaries and areas, points designate locations for activity.

Checkpoints may mark routes or other places that require a reference. In space operations, for example, checkpoints could mark the point in space where a satellite conducts a burn to transfer orbit (a general action point) or where a radar sensor might look to begin its search. Related control measures include the linkup point (Row 3), where a ground segment force links up with its gaining headquarters, and the fly-to-point (Row 4), which may be useful for depicting a GEO satellite's final orbital location. These points are designated simply by number—for example, "Fly-to-Point 4." Additional icons may be placed inside the checkpoint, particularly when depicting supply actions. As in Row 2, the checkpoint symbol has the symbol for Class III (petroleum products) inside of it, denoting that this is a location where units can go to receive additional fuel. In space, a future refueling satellite may fulfill the function of a fuel truck, meeting satellites that require additional fuel at that location. Waypoints (Row 5) may mark points along the necessary routes, but unlike checkpoints—which could be used to accomplish the same function—waypoints, by definition, must be loaded into a navigational aid system, a particularly intriguing thought because it would require satellites with sufficient artificial intelligence to maneuver themselves.

Row	Description	Notes/Reference	Example Usage
1	Checkpoint	MIL-STD-2525D, Appendix H—Command and Control Points, Table H-VI, p. 402	
2	Checkpoint (Class III)	MIL-STD-2525D, Appendix H—Command and Control Points, Table H-VI, p. 402	
3	Linkup Point	MIL-STD-2525D, Appendix H—Command and Control Points, Table H-VI, p. 405	

Row	Description	Notes/Reference	Example Usage
4	Fly-to-Point	MIL-STD-2525D, Appendix H—Command and Control Points, Table H-VI, p. 405	211041DEC09-211841JUN10 FTP 3
5	Waypoint	MIL-STD-2525D, Appendix H—Command and Control Points, Table H-VI, p. 409	X 8
6	Point of Interest—Launch Event	MIL-STD-2525D, Appendix H—Command and Control Points, Table H-VI, p. 406	LE
7	Contact Point	MIL-STD-2525D, Appendix H—Command and Control Points, Table H-VI, p. 402	1
8	Coordinating Point	MIL-STD-2525D, Appendix H—Command and Control Points, Table H-VI, p. 403	⊗

Row	Description	Notes/Reference	Example Usage
9	Special Point	MIL-STD-2525D, Appendix H—Command and Control Points, Table H-VI, p. 408	
10	Point of Interest	MIL-STD-2525D, Appendix H—Command and Control Points, Table H-VI, p. 406	
11	Start Point	MIL-STD-2525D, Appendix H—Command and Control Points, Table H-VI, p. 408	
12	Choke Point	Suggested	

Table 18: Points Useful for Space Operations. Source: Table compiled by authors with information modified from MIL-STD-2525D, Appendix H.

Points of interest are a similar graphic to checkpoints, but they serve to mark sites that are not necessarily part of the operational plan. For example, a China-operated ground station in South America may be noteworthy, but it would likely not affect operational mission planning for a launch and orbital insertion out of Florida. Significantly, however, a specific symbol exists called "Point of Interest—Launch Event" to mark launches (Row 6).

In addition to checkpoints and points of interest, another type of point is the contact point (Row 7), traditionally used in land warfare to mark a spot on the map where "two or more units are required to make contact."* This control differs from a linkup point in its stringency; its use means that units must make contact there before operations can continue. For space operations, such a point could denote where the unit responsible for launch makes contact with the unit responsible for beginning orbital operations. In a coordinated attack from the ground, a mobile jammer makes contact with a ground-based antisatellite laser weapon at an established contact point. The laser blinds the target satellite to deny its ability to image the location of the jammer (and other friendly forces). Meanwhile, the jammer denies the enemy ground control site from being able to access the target satellite, thus thwarting any efforts by controllers to counter the laser attack. Alternatively, as in in the deorbiting and hypersonic glide vehicle scenarios, contact points may be established at necessary locations along the borders between combatant commands. Indeed, in land warfare, contact points are often superimposed on boundary graphics near road intersections or other convenient meeting places.

Similar to the contact point is the coordinating point (Row 8). Coordinating points also often happen along boundaries, but their intent can be accomplished without such a deliberate braking of operational tempo—by radio contact, for example, rather than by physical contact. In the EM spectrum, a coordination point may occur on a certain frequency where two units are conducting mutually supporting operations.

Potential uses for points in space warfare do not stop there, however. Critical "choke points" or key portions of orbital regimes (i.e., geosynchronous orbits supporting a current battlefield, launch corridors for military satellites, or antipodal points on the opposite sides of the earth from launch sites over which all nonmaneuvered missile launches fly) may require graphical designation—perhaps as a special point (Row 9) or generic point of interest (Row 10). Additionally, the point from which a satellite initiates its maneuver may deserve marking (a start point, Row 11). If this start

---

* MIL-STD-2525D, Appendix H—Command and Control Points, Table H-VI, 402.

point exists within a specific combination of orbital parameters (altitudes, inclinations, eccentricities, and/or right ascensions of the ascending node) that make it an optimal jumping-off point for simultaneous attacks against key satellite targets, we may classify this location as a choke point (Row 12). While points in tellurian domains denote precise locations, there is admittedly a bit of orbital "wiggle room" in defining these points. The locations of these points are actually changing with time in orbital space, but they still share key orbital parameters, such as altitude, inclination, and eccentricity. The sample icon in Row 12 specifies the altitude of the choke point (3,000 km) and the inclination (98.8 degrees). With the right visualization software (see Figure 62: SAW—All Altitudes) a user can visualize additional choke points as "fixed" in space.

Thinking about military space operations requires thinking about three distinct but interrelated operating areas (OA) for every operation: a surface OA, an orbital OA, and an electromagnetic OA. Existing symbology for lines and areas provide the concepts by which to divide the largest assigned OA, the area of responsibility. As noted, although the orbital segment of space operations occurs in space command's AOR, day-to-day operations require the involvement of global forces operating across wide swaths of the electromagnetic spectrum. It is therefore necessary to consider the interrelation of the boundaries and areas vis-à-vis multiple boundary lines and areas.

While the use of areas of operations (AOs) needs little explanation for ground-based operations, their expansion into space required consideration of orbital motion and mission needs. Four organizing principles emerged to divide up the AOR, each with its own set of strengths and weaknesses. Whether establishing AOs in space based on geographic projection, mission responsibility, orbital segment, or altitude-inclination relations, some combination of the four methods will be necessary for the execution of any operational-level plan. Furthermore, treating the electromagnetic spectrum like an AO ensures that it is considered as its own entity within the planning process with its own NAIs and TAIs—a particularly useful technique, especially when coordinating for the employment of cyber capabilities.

Within the symbolic lexicon, command and control lines, areas, and points are vital to command and control of forces, but they are only a small part of the entire lexicon. They establish a foundation upon which to employ a myriad of additional symbols to cover the activities of all other warfighting functions and begin to suggest the C2 relationships, infrastructure, and sequencing that will be necessary to accomplish the objectives of the operation. When combined with other symbols and reassessed throughout the planning process, they help ensure the development and visualization of a feasible, suitable, and acceptable plan.

Section 4: Maneuver Control Measures

With a knowledge of unit symbols and symbols for command and control in hand, attention now may turn to the employment of these units through maneuver. Maneuver control measures regulate the maneuver warfighting function, and although maneuver requires the combination of movement and fires in relation to the enemy, the movement and fires aspects of symbology will be treated in separate sections. There is, as a point of note, no maneuver control symbology specific to the space domain. This limitation, however, presents a minor problem. As with other types of symbology, maneuver control symbology for space operations may draw liberally from the existing maneuver symbology of the land, air, and maritime domains—a beneficial consequence in the quest to create commonality of language across domain-focused practitioners. In the employment of such symbols, it is important to consider orbital motion as a component of maneuver, to discuss, as in the previous chapter, the implications of considering the electromagnetic spectrum as an arena for maneuver, and to build these measures into ground segment operations.

Orbital Motion

To plan for orbital assets, it is necessary to know their positions and motions, particularly in relation to other assets of interest. For this purpose, dynamic graphic amplifiers are useful to denote an area of uncertainty (AOU) around the asset and with reference to its projected path. For a satellite, the "area of uncertainty" is misleading.* The more appropriate term for orbital operations is *volume of uncertainty* (VOU)—a term that we coin here to highlight the three-dimensional nature of the problem. In air domain symbology, an AOU may be depicted as an ellipse, a bearing box, a line of bearing AOU without error lines, or a line of bearing with error lines (see Figure 31). In the space domain, each VOU becomes the three-dimensional expansion of the two-dimensional AOU. Thus, the ellipse AUO becomes an ellipsoid VOU, the bearing box AUO a rectangular prism, and the bearing AOU with error lines becomes a cone, the shell of which represents the greatest possibility of error in the orbital path. As with AOUs, the accuracy of these VOUs is a function of the sensitivity of the sensor collecting positioning data on the object and the time since the sensor's last observation. Since satellites drift over time because of solar pressure and small differences in gravity across the earth and since ground-based radars or optical sensors are only so accurate, the VOU is likely to increase with time—perhaps to a point where prudence dictates either a sensor revisit or a decision that available intelligence is simply too stale to be useful.

The line of bearing amplifiers may be particularly useful to detect off-nominal satellite bearings—those bearings that have unexpectedly deviated from the projected course since the last measurement was taken and may indicate harmful intent. Variation between a "dead reckoning" (DR) measurement that projected the motion forward in time and the updated bearing could indicate a deliberate maneuver, which also may be an indicator of adversarial intent. The "furthest-on circle DR trailer," when transformed into a sphere, can be used to assess the possible distance a satellite could have moved between measurements. Finally, the pairing line amplifier—like the line of bearing, it is useful as a simple line—"connects two objects and is updated dynamically as the positions of the two objects change." This functionality

---

*   The authors realize that we did not insist upon referring to areas of operation as volumes of operation in the earlier sections. However, we insist upon *volume of uncertainty* here to avoid suggesting that the satellite is merely situated inside of a disc on an elliptical plane. *Volume* implies that the satellite is located inside of an ellipsoid, accounting for the uncertainty in three dimensions at any moment in time.

would be useful in assessing rendezvous and proximity operations between satellites or between a kinetic kill vehicle and a satellite.

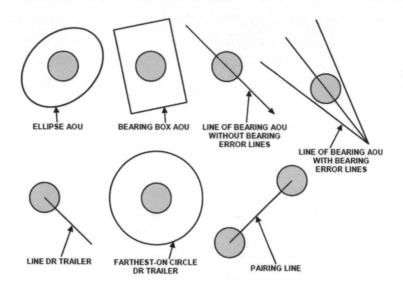

Figure 31: Some Dynamic Graphics Amplifiers for Icon-Based Symbols. *Source:* Modified from MILSTD-2525D, p. 29.

Figure 32 shows how a line of bearing amplifiers may be used in a common operating picture display. In this scenario, a geosynchronous electronic warfare satellite is bearing toward a target communications satellite. The graphic includes additional information about the expected rendezvous time (time on target) and the friendly satellite's visual brightness. Brightness expresses the visibility to sensor systems. Dimmer objects are less susceptible to tracking by optical sensors. Likewise, a dimmer satellite—as viewed by the target—would be less likely to be detected by the target, thereby reducing reaction time.

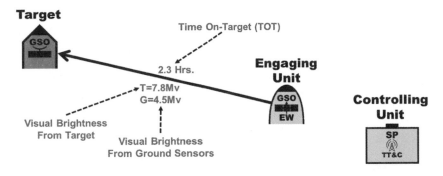

Figure 32: Engagement Control Scenario with Satellites. *Source:* Created by Paul Szymanski.

Beam Patterns and Radar Fans

A second aspect of using symbology to better understand the space operating environment requires consideration of how to portray electromagnetic
manifestations, particularly radio frequency beam patterns and radar fans.
These symbols may traditionally fit within a discussion of control measures
for the fires warfighting function. However, they are introduced here for
two reasons. First, fires are a part of maneuver, and if one wished to view
electromagnetic emissions as fires, one would be in good company doing
so. Second, and less orthodox, is the idea that movement is continuously
occurring inside the portions of the electromagnetic spectrum, and as a
result, the electromagnetic spectrum is an arena for maneuver. The implication is that space operations forces must maneuver in all segments—orbital,
electromagnetic, and surface—and the maneuvers must be deliberately
planned and executed in coordination with each other.

Spectral movement may occur in three ways: movement internal to the
signal, movement of the signal itself, and movement of groups of signals.
One may think of movement internal to the signal as an operator-level
manipulation of the signal's parameters, for example modulation or polarization changes. The movement of the signal itself happens when operators
move from signal to signal, channel to channel, transponder to transponder,
satellite to satellite, or even satellite to other means (e.g., fiber, telephone
line, or radio relay). Historically, this switching has often been transparent

to the user, who, when everything is going well, is hardly aware of the frequency-hopping algorithm or the on-board activities of a satellite controller half a world away. Finally, radio frequency movement can happen through beam shaping or through movement of a terminal or the satellite itself, and this is the most practical case requiring graphic control measures.

Satellites uplink or downlink radio signals in various beam patterns. These beam patterns are typically projected on the earth's surface, forming a footprint that is often visualized as a pattern of contour lines (see Figure 33). Though this is the convention, there is no formal control measure for a satellite communication beam pattern, but there is a similar symbol in the radiation dose rate contour line. In that control measure, the various contour lines correspond to various doses of radiation (in centigrays or cGy), the dosage dissipating as the lines extend farther from the center (Row 1). Similarly, satellite communications signal strength contour lines would each correspond to the signal strength (in decibels or dB) of the signal over a given area. Table 19 shows the established radiation dose rate contour line control measure as a model for the proposed satellite communications signal contour line control measure. Interestingly, if extrapolated into a three-dimensional shape, the radiation dose contour line control measure may have operational applicability in depicting the radiation environment in space, particularly after an orbital nuclear detonation, or in visualizing the energy pattern of a terrestrial or space-based high-power microwave (HPM) weapon.

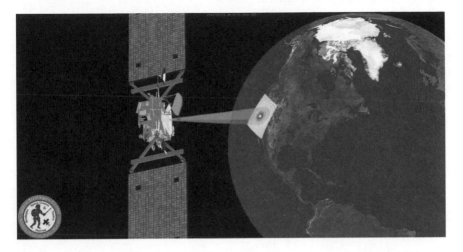

Figure 33: A Notional Satellite Beam Signal Strength Pattern Projected onto the Earth and Visualized as Contour Lines. *Source:* Graphic created by Paul Szymanski using the Satellite Orbit Analysis Program (SOAP) from The Aerospace Corporation.

Row	Description	Notes/Reference	Icon
1	Radiation Dose Rate Contour Line	MIL-STD-2525D, Appendix H—Control Measure Symbols, Table HXXI, p. 609	30 cGy 100 cGy 300.cGy
2	Satellite Communications Signal Contour Line	Proposed	3dB 4dB 5dB

Row	Description	Notes/Reference	Icon
3	Weapon/ Sensor Range Fan, Circular	MIL-STD-2525D, Appendix H—Control Measure Symbols, Table HXVIII, p. 566	
4	Weapon/ Sensor Range Fan, Sector	MIL-STD-2525D, Appendix H—Control Measure Symbols, Table HXVIII, p. 567	

Table 19: Symbology for Beam Patterns and Radar Range Fans. Source: Table compiled by authors with information modified from MIL-STD-2525D, Appendix H.

Ground-based radars used for space domain awareness applications (such as allowing us to refresh our VOUs) require a conceptually similar symbol. Rather than projecting a satellite's radio waves onto the earth, however, radars effectively project their electromagnetic radiation onto the sky. A two-dimensional contour plot projected onto the sky may be useful—if, for example, the radar energy is directed at a single orbital altitude, say 409km, the approximate altitude of the International Space Station. Another application would be to project the maximum detectable ranges of terrestrial-based radars to determine the extent of their ability to observe deep space objects.

In fact, the "weapon/sensor range circular" (Row 3) and "weapon/sensor range fan, sector" (Row 4) control measures accomplish this function for ground-based radars—either omnidirectional or directional radars, respectively. Both control measures are projections on the earth's surface that account for the vertical component of the radar's operation through annotation on the two-dimensional symbol. The circular ring annotations indicate a minimum range ("MIN RG"—in the table example, 1200 meters), and two options of maximum range. The altitude of the unit of employment, which would be depicted as a unit icon in the center of the ring, is ground level ("ALT GL"). The range fan graphic works similarly. Row 4 depicts this graphic with a friendly radar unit at its base, the central arrow denoting the center point of the radar's orientation and the numbers at the edges of the fans denoting the angular distance from the center line.

While these simple visualizations are useful, especially for operational planning, they may not contain the fidelity necessary for a technical analysis. Electromagnetic patterns, whether for satellite communications or for radar beams, are, in real life, neither circular nor neatly contained by smooth radar fan lines. As in the case of VOUs, three-dimensional expansions of the concepts allow for more effective visualization of space operations when necessary. For space situational awareness radars, a three-dimensional radar visualization may show a main beam (as in Figure 34), a main beam with side lobes, or a radar dome or radar sector with a vertical contour—a "layer cake" visualization to depict the amount of energy reaching a particular orbital regime.

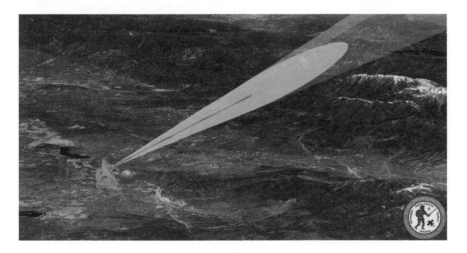

Figure 34: Notional Satellite Ground Station Emitter with Main Lobe Visualization. *Source:* Graphic created by Paul Szymanski using the Satellite Orbit Analysis Program (SOAP) from The Aerospace Corporation.

Maneuver Areas

Maneuver areas work like command-and-control areas and can be equally useful in the employment of ground or space segment forces. In the ground segment usage, their employment follows the traditional usage—a flat outline on a two-dimensional map. For the space segment—as with command-and-control areas—friendly or enemy maneuver areas are properly three-dimensional volumes. Table 20 shows some common maneuver areas. Again, occupied areas are outlined in solid boundaries (Rows 1 and 3), and "planned" or "on-order" areas for friendly forces and suspected enemy force areas are depicted with dashed lines (Rows 2 and 4). Enemy forces can either be marked "ENY" or colored red as in (Rows 3 and 4). Row 5 depicts an assembly area—an "area in which a command is assembled preparatory to further action." One might imagine satellites congregating in advantageous locations (i.e., choke points) in space prior to disbursing for coordinated activities. The final maneuver area that may be of use is the landing zone (LZ). With recoverable spacecraft like the X-37B and with recoverable launch vehicles like the first stage of Space-X's Falcon 9, it may in the future be necessary to designate landing zones and the forces necessary to secure both the zone and the landing craft. Row 6 shows the graphic for a landing zone called "LZ Silver."

Row	Description	Notes/Reference	Icon
1	Friendly Area	MIL-STD-2525D, Appendix H—Control Measure Symbols, Table H-VII, p. 414	
2	Friendly Planned or On-Order Area	MIL-STD-2525D, Appendix H—Control Measure Symbols, Table H-VII, p. 415	

Row	Description	Notes/Reference	Icon
3	Enemy Known or Confirmed Area	MIL-STD-2525D, Appendix H—Control Measure Symbols, Table H-VII, p. 415	ENY ENY
4	Enemy Suspected Area	MIL-STD-2525D, Appendix H—Control Measure Symbols, Table H-VII, p. 415	ENY ENY
5	Assembly Area	MIL-STD-2525D, Appendix H—Control Measure Symbols, Table H-VII, p. 416	AA BLUE
6	Landing Zone (LZ)	MIL-STD-2525D, Appendix H—Control Measure Symbols, Table H-VII, p. 418	LZ SILVER

Table 20: Maneuver Area Symbology. Source: Table compiled by authors with information modified from MIL-STD-2525D, Appendix H.

Defensive Control Measure Symbols

Defensive control measure symbols (Table 21) are a subset of maneuver control measure symbols and include battle positions (Row 1), engagement areas (Row 2), and a variety of observation post symbols (Rows 3–4).* In land warfare, a battle position is "a defensive location oriented on a likely enemy avenue of approach."† One may imagine a platoon of four tanks arrayed across a road down which the enemy is expected to proceed. This platoon is in battle position "XRAY" (Row 1). Multiple platoons occupy multiple battle positions to cover an engagement area ("EA ROCK" in Row 2), the location where they will engage the enemy as it advances. In Row 2, the black portion of the graphic shows the symbol; platoon battle positions are shown in gray for context and illustrate the idea of overlapping fields of fire. The analogue of battle positions with overlapping fields of fire is useful for space electronic warfare, combined arms attacks against satellites, or coordinated radar coverage.

Observation posts sit forward and to the flanks of the friendly positions to gather intelligence on enemy activity and to provide early warning. Although there are a variety of modifiers that one may apply to the generic outpost symbol, symbols for a reconnaissance outpost (Row 3) and a sensor outpost/listening post (Row 4) likely have the most utility for space operations. One may imagine, for example, depicting an array of sensor outposts to show multiple types of different sensors contributing to the same observation effort. Outposts employ target reference points (Row 5) as signposts to direct fires against preplanned locations.‡

Row Number	Description	Notes/Reference	Example
1	Battle Position	MIL-STD-2525D, Appendix H— Control Measure Symbols, Table HVIII, p. 420	XRAY

---

* MILSTD 2525D also includes the symbols for *contain* and *retain* in the section on defensive maneuver control measures. We have chosen to include those in a separate section on tactical tasks.
† 2525D (2014), 420.
‡

Row Number	Description	Notes/Reference	Example
2	Engagement Area (EA)	MIL-STD-2525D, Appendix H—Control Measure Symbols, Table HVIII, p. 424	
3	Recon-naissance Outpost	MIL-STD-2525D, Appendix H—Control Measure Symbols, Table HIX, p. 425	
4	Sensor Outpost/ Listening Post	MIL-STD-2525D, Appendix H—Control Measure Symbols, Table HIX, p. 426	
5	Target Reference Point	MIL-STD-2525D, Appendix H—Control Measure Symbols, Table HIX, p. 427	361

Table 21: Some Defensive Control Measure Symbols. Source: Table compiled by authors with information modified from MIL-STD-2525D, Appendix H.

Offensive Control Measure Symbols

While the dynamics graphics amplifiers of Figure 31 depict a spacecraft's movement, these symbols fall short of depicting maneuver in relation to an enemy. For that function, we require offensive control measure symbols (Table 22). These symbols are the arrows on the map that depict the main attack (Row 1), a supporting attack (actual, Row 2; planned or on order, Row 3), or a feint (Row 4).

Row	Description	Notes/Reference	Symbol
1	Main Attack	MIL-STD-2525D, Appendix H—Offensive Control Measure Symbols, Table H-X, p. 429	
2	Supporting Attack	MIL-STD-2525D, Appendix H—Offensive Control Measure Symbols, Table H-X, p. 430	
3	Supporting Attack, Planned or On Order	MIL-STD-2525D, Appendix H—Offensive Control Measure Symbols, Table H-X, p. 431	
4	Axis of Advance for a Feint	MIL-STD-2525D, Appendix H—Offensive Control Measure Symbols, Table H-X, p. 431	

Table 22: Some Offensive Control Measure Symbols. Source: Table compiled by authors with information modified from MIL-STD-2525D, Appendix H.

Direction of attack symbols (Table 23) perform similarly to the control measures shown in Table 22 but with an important difference is specificity. The "direction of main attack" series of arrows direct a "specific direction or route that the main attack or center of mass of the unit will follow."* The search area/reconnaissance area graphic (Row 6) does not show a direction of attack, per se, but it does specify the direction of reconnaissance. This symbol, as with the others in this section, takes on unique implications when considered through the lens of space warfare.

Row	Description	Notes/Reference	Example
1	Friendly Direction of Main Attack	MIL-STD-2525D, Appendix H—Offensive Control Measure Symbols, Table H-XI, p. 431	
2	Enemy Direction of Main Attack	MIL-STD-2525D, Appendix H—Offensive Control Measure Symbols, Table H-XI, p. 433	ENY
3	Friendly Direction of Supporting Attack	MIL-STD-2525D, Appendix H—Offensive Control Measure Symbols, Table H-XI, p. 433	
5	Direction of Attack for a Feint	MIL-STD-2525D, Appendix H—Offensive Control Measure Symbols, Table H-XI, p. 433	

---

*   2525D (2014), 432.

Row	Description	Notes/Reference	Example
6	Search Area/Recon-naissance Area	MIL-STD-2525D, Appendix H—Offensive Control Measure Symbols, Table H-XI, p. 444. Shown with a space unit symbol for context.	

Table 23: Direction of Attack, Search Area/Reconnaissance Symbols. Source: Table compiled by authors with information modified from MIL-STD-2525D, Appendix H.

The primary value of the symbols of Table 22 and Table 23 rests in their use in conceptual planning. On the ground, the infantry company will follow the path of the arrow as it is drawn atop the map; the interpretation of the arrow is literal. Figure 35 shows an enemy satellite with a kinetic kill vehicle as the main attack against a friendly satellite. It also shows a laser attack ASAT feinting toward another geosynchronous satellite. Conceptually, this is straight-forward, and one may depict such a graphic of a two-dimensional world map (as appears in the next chapter), perhaps annotating the time of the anticipated attack and the orbital characteristics of the satellites. To accomplish such an attack, however, may require several orbits and multiple vector changes over a period of days. Furthermore, the existence of a main attack implies the existence of one or more supporting attacks. The calculations to accomplish such maneuvers require the expertise of the orbitologist, but coordinating the tactical action and understanding the military implications thereof are squarely the responsibility of the space warfighter. A more complete picture of the tactical option for space, however, requires a consideration of tactical tasks, to which we now turn.

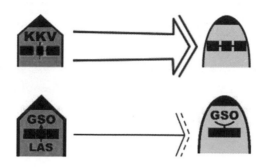

Figure 35: Enemy Kinetic Kill Vehicle as the Main Attack; Laser
Attack as a Feint. *Source:* Created by authors.

Tactical Mission Tasks

Tactical mission task symbols are a subset of maneuver control measure symbols. Like the symbols already discussed in this chapter, tactical tasks provide specific guidance to subordinate commanders, explaining the missions they must accomplish as part of the larger operation. Thus, what makes tactical tasks more significant than other control measures is that they reflect the purpose of the unit—what the unit does when it gets to the end of the arrow. If assembly areas and attack direction arrows set the stage, the tactical task is the purpose of the force—*how* the force achieves success. Ground forces most commonly use the following symbols—as evidenced by their doctrinal definitions—but as with previous symbols, their applicability transcends any single domain.

The following table (Table 24) lists the tactical tasks identified for ground force operations. Rather than trying to amend accepted doctrinal definitions, the definitions within the table intentionally reflect the ground-force usage. Again, we confront our language problem. Some definitions that may be entirely suitable for an army's use do not quite seem to fit the needs of a space operations force, yet one must consider each in turn and ask why such differences exist and how common understanding may be achieved despite these differences. To aid in this consideration, the example usage column places the tactical task symbol (in black ink) within the context of interacting spacecraft. These symbols, of course, pertain to the ground segment of space operations forces, and many may be applied to space operations activity within the electromagnetic spectrum as well. Furthermore, although the examples typically only involve one or two satellites, it would be possible to employ multiple satellites for the same purpose and combined arms attacks employing different phenomenology weapon systems from different directions and timelines.

Row	Task/Definition	Example Use
1	**Attack by Fire**: A tactical mission task in which a commander uses direct fires, supported by indirect fires, to engage an enemy force without closing with the enemy to destroy, suppress, fix, or deceive that enemy (ADRP 1-02, 1-7). Space example: employing space EW to fix a satellite in place while directed energy weapons are brought to bear on a target.	
2	**Block**: A tactical mission task that denies the enemy access to an area or prevents his advance in a direction or along an avenue of approach (ADRP 1-02, 1-9). Space examples: deploying multiple ASATs covering a particular orbital regime to prevent an enemy from accessing it. Other more radical techniques may include purposely creating debris fields or increasing radiation belts to deny an adversary use of that orbital region of space.	
3	**Bypass:** A tactical mission task in which the commander directs his unit to maneuver around an obstacle, position, or enemy force to maintain the momentum of the operation while deliberately avoiding combat with an enemy force (ADRP 1-02, 1-11). Space example: taking the long way around a particular orbit to avoid contact or observation, even though this may take additional orbital fuel (delta-v).	

Row	Task/Definition	Example Use
4	**Canalize:** A tactical mission task in which the commander restricts enemy movement to a narrow zone by exploiting terrain coupled with the use of obstacles, fires, or friendly maneuver (ADRP 1-02, 1-12). Space example: employing unusual orbital dynamics to "tunnel" his adversary's space systems into regions of the friendly commander's choosing. One may force an adversary into certain orbits by heavily defending other orbits or by pumping up radiation belts or debris clouds of some orbits. The locations and ranges of friendly space sensors would influence an adversary to employ certain orbits to avoid detection.	
5	**Contain:** A tactical mission task that requires the commander to stop, hold, or surround enemy forces or to cause them to center their activity on a given front and prevent them from withdrawing any part of their forces for use elsewhere (ADRP 1-02, 1-20). Space example: concentrating ASAT forces or creating a fake high-value target so as to direct adversary forces in a way that is advantageous to friendly goals.	
6	**Counterattack:** Attack by part or all of a defending force against an enemy attacking force, for such specific purposes as regaining ground lost or cutting off or destroying enemy advance units, and with the general objective of denying to the enemy the attainment of the enemy's purpose in attacking (ADRP 1-02, 1-22). Space example: when enemy ASATs are converging on a high-value region of geosynchronous space, redirecting friendly ASATs from beyond GEO to attack the attackers.	

Row	Task/Definition	Example Use
7	**Counterattack by Fire:** Attack [by fire] by part or all of a defending force against an enemy attacking force, for such specific purposes as regaining ground lost, or cutting off or destroying enemy advance units, and with the general objective of denying to the enemy the attainment of the enemy's purpose in attacking. In sustained defensive operations, it is undertaken to restore the battle position and is directed at limited objectives (ADRP 1-02, 1-22). Space example: use of friendly EW, HPM, or directed energy weapons to shock and confuse adversary space attack maneuvers and preparations.	
8	**Delay:** To slow the time of arrival of enemy forces or capabilities or alter the ability of the enemy or adversary to project forces or capabilities (ADRP 1-02, 1-27). Space example: frustrate adversary attack preparations by denying his command and control (along with space surveillance) of his space systems, including his ability to launch replacement satellites or scatter his mobile terrestrial space support assets.	
9	**Demonstration:** In military deception, a show of force in an area where a decision is not sought that is made to deceive an adversary. It is similar to a feint but no actual contact with the adversary is intended (ADRP 1-02, 1-27). Space example: threaten adversary space assets by sending inspection satellites while simultaneously maneuvering friendly ASATs close to adversary targets. These threats can be against targets that actually make no sense in the current military/political environment so as to confuse him about your ultimate objectives.	
10	**Destroy:** A tactical mission task that physically renders an enemy force combat-ineffective until it is reconstituted. Alternatively, to destroy a combat system is to damage it so badly that it cannot perform any function or be restored to a usable condition without being entirely rebuilt (ADRP 1-02, 1-28). Space example: there is a multitude of ways to destroy a satellite's ability to conduct its mission that do not involve debris generation. Cyberattacks may be the most promising.	

Row	Task/Definition	Example Use
11	**Disengage:** A tactical mission task where a commander has his unit break contact with the enemy to allow the conduct of another mission or to avoid decisive engagement (ADRP 1-02, 1-29). Space example: ASATs do not have the intended effects, so the friendly commander breaks off his main attack against an adversary satellite constellation and attempts to achieve mission goals through other means like cyber satellite network attacks.	DIS
12	**Disrupt:** A tactical mission task in which a commander integrates direct and indirect fires, terrain, and obstacles to upset an enemy's formation or tempo, interrupt his time-table, or cause enemy forces to commit prematurely or attack in piecemeal fashion (ADRP 1-02, 1-30). Space example: friendly commander employs multiple phenomenology ASATs, such as KKVs, lasers, and cyber weapons to influence the adversary to maneuver his satellites, disrupting his space missions in doing so. This action should also take into consideration some no-go regions of space with high radiation belts or debris clouds.	
13	**Envelopment:** A form of maneuver in which an attacking force seeks to avoid the principal enemy defenses by seizing objectives behind those defenses that allow the targeted enemy force to be destroyed in their current positions (ADRP 1-02, 1-33). Space example: seizing translunar space to attack geosynchronous satellites coming in from an unusual direction. Like a frontal attack, an envelopment could involve "Hun out of the sun" attacks—approaching an adversary satellite with the sun behind you so as to blind its defensive sensors.	E
14	**Exploitation:** An offensive operation that usually follows a successful attack and is designed to disorganize the enemy in depth. Space example: after an attack on an adversary's space command and control systems, take advantage of his confusion and inability to command satellites and attack his space systems with relative impunity" (ADRP 1-02, 1-35).	

Row	Task/Definition	Example Use
15	**Exfiltration:** The removal of personnel or units from areas under enemy control by stealth, deception, surprise, or clandestine means (ADRP 1-02, 1-35). Space example: removal of personnel from critical terrestrial space systems before an impending attack. In space, maneuvering of threatened satellites beyond the reach of an adversary's ASAT and targeting sensors.	
16	**Feint:** In military deception, an offensive action involving contact with the adversary conducted for the purpose of deceiving the adversary as to the location and/or time of the actual main offensive action (ADRP 1-02, 1-36). Space example: visibly concentrating friendly ASATs against an adversary target to draw attention away from covert weapon system attacks on the other side of the earth.	
17	**Fix:** A tactical mission task where a commander prevents the enemy from moving any part of his force from a specific location for a specific period. Fix is also an obstacle effect that focuses fire planning and obstacle effort to slow an attacker's movement within a specified area, normally an engagement area (ADRP 1-02, 1-39). Space example: using cyberattack means to "fix" an adversary satellite in place so the enemy cannot command it into defensive modes or maneuver it into another region. One might use a similar technique to fix launch vehicles, denying an adversary's ability to launch replacement satellites. Note that as a tactical symbol, the zigzag arrow has a specific meaning. Its use to denote directed energy attacks should be avoided unless the intent of the attack is to fix the enemy.	

Row	Task/Definition	Example Use
18	**Follow and Assume:** A tactical mission task in which a second committed force follows a force conducting an offensive task and is prepared to continue the mission if the lead force is fixed, attritted, or unable to continue (ADRP 1-02, 1-39). Space example: sending two ASATs against one target and having the second ASAT hold back to see if another attack is necessary to increase probability of kill. Also, purposely exploding an apparently harmless "civil" space system after several ASATs have exited the main bus while flying over Antarctica to avoid detection. Have these multiple ASATs hide out in the resultant debris cloud.	
19	**Interdict:** A tactical mission task where the commander prevents, disrupts, or delays the enemy's use of an area or route (ADRP 1-02, 1-50). Space example: employing cyberattack means to disrupt satellite terrestrial launches and redeployments of on-orbit assets. Also, one could use increased radiation belts and debris clouds to prevent the use of certain orbits, along with deploying ASAT systems to cover other critical orbits.	
20	**Isolate:** A tactical mission task that requires a unit to seal off—both physically and psychologically—an enemy from sources of support, deny the enemy freedom of movement, and prevent the isolated enemy force from having contact with other enemy forces (ADRP 1-02, 1-51). Space example: employing cyber and direct destruction forces to sever space systems from terrestrial controllers and command elements. Can also attempt to deny satellite-to-satellite in-space communications systems like relay satellites, which are critical choke points for control and downloading mission data.	
21	**Neutralize:** A tactical mission task that results in rendering enemy personnel or materiel incapable of interfering with a particular operation (ADRP 1-02, 1-65). Space example: typical ASAT techniques of KKV, directed energy, cyber, direct attacks on ground systems, and so on. Neutralizing falls short of destroying and may not be as long-lasting.	

Row	Task/Definition	Example Use
22	**Occupy:** A tactical mission task that involves a force moving a friendly force into an area so that it can control that area. Both the force's movement to and occupation of the area occur without enemy opposition (ADRP 1-02, 1-67). Space example: maneuvering to and occupying key orbital regimes that are covered by multiple ASAT systems. Best if done preconflict.	
23	**Passage of Lines (Forward):** An operation in which a force moves forward [. . .] through another force's combat positions with the intention of moving into or out of contact with the enemy (ADRP 1-02, 1-69). Space example: when hostilities escalate, an ASAT passes forward of an inspector satellite that was holding a target under observation. Analogous to a tank battalion passing through its cavalry screen.	
24	**Passage of Lines (Rearward):** An operation in which a force moves [. . .] rearward through another force's combat positions with the intention of moving into or out of contact with the enemy (ADRP 1-02, 1-69). Space example: friendly ASAT forces must maneuver for replenishment (refueling, recharging, new missiles, and so on) and tactical relocation to protect another region of space or another potential corridor of adversary ASAT attacks. In so doing, they regress through the position of another ASAT force.	
25	**Penetration:** A form of maneuver in which an attacking force seeks to rupture enemy defenses on a narrow front to disrupt the defensive system (ADRP 1-02, 1-70). Space example: punching through adversary ASAT defenses by using high-speed or unusual orbital maneuvers to gain access to a highly defended critical asset behind defensive systems.	

Row	Task/Definition	Example Use
26	**Relief in Place:** An operation in which, by direction of higher authority, all or part of a unit is replaced in an area by the incoming unit, and the responsibilities of the replaced elements for the mission and the assigned zone of operations are transferred to the incoming unit (ADRP 1-02, 1-77). Space example: New ASATs occupy friendly orbital positions that are already in use to allow friendly ASAT forces to maneuver for replenishment (refueling, recharging, new missiles) or for tactical relocation.	
27	**Retain:** A tactical mission task in which the commander ensures that a terrain feature controlled by a friendly force remains free of enemy occupation or use (ADRP 1-02, 1-79). Space example: making sure critical space orbital choke points are continuously and vigorously controlled by friendly space assets.	
28	**Retirement:** A form of retrograde in which a force out of contact moves away from the enemy (ADRP 1-02, 1-79). Space example: friendly ASAT forces maneuvering to a new location, possibly because the threat against which they were positioning is no longer significant or possibly because an in-contact unit needs to withdraw, which requires the out-of-contact unit to retire.	
29	**Secure:** A tactical mission task that involves preventing a unit, facility, or geographical location from being damaged or destroyed as a result of enemy action (ADRP 1-02, 1-81). Space example: defensive actions to ensure key terrestrial space-related assets are defended, such as command and control locations, space surveillance sites, tracking systems, launch sites, rocket fuel production and storage sites, and terrestrial-based space weapon systems, such as direct ascent ASAT missiles, lasers, cyber antennas, and airports used by air-launched ASATs.	

Row	Task/Definition	Example Use
30	**Security (Screen):** A security task that primarily provides early warning to the protected force (ADRP 1-02, 1-80). Space example: redeployment of mobile space surveillance assets (including SIGINT) to cover critical theaters of operation. Also includes fixed space surveillance sensors that can be reconfigured to emphasize observations of certain critical adversary satellite systems. A group of inspector satellites could screen but not guard.	
31	**Security (Guard):** A security task to protect the main force by fighting to gain time while also observing and reporting information and preventing enemy ground observation of and direct fire against the main body. Units conducting a guard mission cannot operate independently because they rely upon fires and functional and multifunctional support assets of the main body (ADRP 1-02, 1-43). Space example: key ASAT attacks that frustrate adversary ASAT systems maneuvering into position to attack main friendly forces. Also includes moving forward friendly space-based surveillance assets and satellite inspector systems. Inspector satellites alone could not guard. Because guarding requires fighting to protect the main force, it would necessarily require ASATs.	
32	**Seize:** A tactical mission task that involves taking possession of a designated area using overwhelming force (ADRP 1-02, 1-81). Space example: taking possession of space orbital choke points that are then continuously and vigorously protected by friendly space assets. By definition, an enemy presence is implied; if an enemy was not present, the task would be an occupation rather than a seizure.	

Row	Task/Definition	Example Use
33	**Support by Fire:** A tactical mission task in which a maneuver force moves to a position where it can engage the enemy by direct fire in support of another maneuvering force (ADRP 1-02, 1-86). Space example: employing supporting fires from lasers and cyber weapons in support of forces maneuvering closer to short-range attack positions.	
34	**Suppress:** A tactical mission task that results in temporary degradation of the performance of a force or weapons system below the level needed to accomplish the mission (ADRP 1-02, 1-87). Space example: Degrading satellite self-detection capabilities, delaminating satellite protective materials, or destabilizing internal satellite subsystems through cyber means. Suppression could also be achieved through a feint (i.e., feint to suppress). This can be as simple as fake attack maneuvers close to a satellite in order to drain its fuel reserves or trick it into exposing war reserve satellite protection modes, such as unnecessary deployment of chaff or misdirection balloons. Satellites usually cannot continue their mission while maneuvering or reconfiguring major subsystems.	
35	**Turn:** A tactical mission task that involves forcing an enemy force from one avenue of approach or mobility corridor to another (ADRP 1-02, 1-94). Space example: employing cyberattack means to disrupt satellite terrestrial launches and redeployments of on-orbit assets while also deploying ASAT systems to cover other critical orbital paths. As on land, a turning movement involves attacking lines of communication to "turn" the enemy out of his position or risk being surrounded.	
36	**Withdraw:** A planned retrograde operation in which a force in contact disengages from an enemy force and moves in a direction away from the enemy (ADRP 1-02, 1-98). Space example: withdrawing friendly space assets from certain orbits. This may be done to confuse the enemy or as an admission of inability to adequately defend these assets.	

Row	Task/Definition	Example Use
37	**Withdraw under Pressure:** "Withdraw under pressure" shares the same definition as "withdraw" but suggests a greater sense of urgency. The tactical task symbol is modified. Space example: withdrawing friendly space assets from certain orbits while having the support of other ASAT forces to defend this withdrawal.	

Table 24: Tactical Tasks. Source: Definitions come from ADRP
1-02. Symbol arrangements created by Jerry Drew.

While many of the task definitions are self-explanatory, the addition of simple examples helps to concretize their meaning in a space warfare context. From this long list of tasks and simple examples, several lend themselves to additional exploration because they draw out key ideas about space operations. Of the thirty-seven tactical tasks here, six tasks are treated in depth. What emerges from this discussion is the notion that space operations are very much a maneuver and fires problem set.

Row	Task/Definition	Example Use
1	**Attack by Fire**: A tactical mission task in which a commander uses direct fires, supported by indirect fires, to engage an enemy force without closing with the enemy to destroy, suppress, fix, or deceive that enemy (ADRP 1-02, 1-7).	

As the definition states, a force conducts an attack by fire with a direct-fire asset supported by indirect fires. Direct fires are those that the individual firing the weapon can also observe. Machine guns and tanks, for example, are direct-fire weapons because gunners can observe their own fires and adjust accordingly. If the operators of a space system, for example a ground-based satellite jammer, are capable of assessing the success or failure of their efforts without external assistance, this jammer is a direct-fire weapons system. Oppositely, indirect fires are those for which the individual firing the weapon cannot observe the effects of their fires; they rely on an external actor to observe for them and to adjust their aim—as in the case of artillery fire and forward observers. A ground-based satellite jammer may, depending upon its concept of operation, rely on another organization to observe and report upon its fires. In addition, directed energy, lasers, or high-power microwave attacks against satellites can have effects that might not be readily visible to the attacking systems and may require close inspection satellites for damage assessment.

While ground-based jammers provide a ready example, given the DIA's assessment of possible future capabilities, several classes of spacecraft may be capable of this tactical task in the near future. Whether through a kinetic attack or through radio-frequency or microwave attacks, one satellite could feasibly attack another by fire. Similarly, multiple satellites could take part in a coordinated attack by fire (possibly with multiphenomenology weapons, approach angles, and time sequencing to keep the adversary off-balance), just as multiple tanks in a tank platoon would be given the platoon-level task to attack by fire. Assuming that the satellite operators are capable of observing their own actions—making such satellites direct-fire weapons systems—such activities would fall under the category of an attack by fire.

Cross-domain options emerge by combining both ground-based and space-based attack-by-fire missions. One may, for example, sequence a radio-frequency attack from an on-orbit asset with a radio-frequency attack from a ground-based jammer, or one may combine a simultaneous microwave attack against an enemy spacecraft's communication suite while a ground-based laser attacks the optical sensor of the same spacecraft. Even if one of these activities falls under the category of indirect fire, the force still achieves the tactical task of *attack by fire*. Any such combination of direct and indirect fires, of ground- and space-based actions, or of attacking multiple satellite subsystems may produce the desired effect of destroying, suppressing, fixing, or deceiving the enemy. For an added twist, an attacking country may combine overt and covert space weapons systems such that the overt system is blamed for the attack while the covert system remains anonymous and not subjected to enemy counterattacks. This will keep the enemy always guessing about the source of the next attack, preserves the anonymous weapon system for employment in future conflicts, and minimizes potentially negative political outcomes of employing sensitive space attack means.

Row	Task/Definition	Example Use
2	**Block**: A tactical mission task that denies the enemy access to an area or prevents his advance in a direction or along an avenue of approach (ADRP 1-02, 1-9).	

In traditional ground-force symbology, the *Attack by Fire* and *Block* graphics depict precisely on a map where a force will conduct its task. This precision allows for the deliberate planning of, among other things, supporting fires. For space operations, however, these graphics take on more of an intent-based function, rather than a precision-depiction function. On orbit, *the intent of blocking*—denying access or preventing desired movement—may be achieved without setting up a blocking position. Importantly, blocking is distinct from fixing, disrupting, or delaying because its aim is permanent, rather than temporary, prevention. Denial of access to an area may imply occupying a valuable orbit or portion of the electromagnetic spectrum so that another satellite or user cannot. Deployment of space weapons covering a critical orbital regime would serve the same purpose of blocking any adversary satellites from deploying into that denied space.

The ways of achieving a block are numerous. In a simple scenario, a spacecraft with a robotic arm may prevent another spacecraft from continuing along its intended course. At the extreme end of options lies a minefield in space or a debris field set to prevent occupation or undesirable orbital motion. Although the most recent DIA assessment does not include space mines as a threat, they are mentioned here because of their historical consideration as a potentially viable option. The use of space mines or a deliberate debris field suggests the second doctrinal meaning of *block*, that of obstacle effect rather than tactical task. In ground warfare, a unit that is given the tactical task of blocking often employs a blocking obstacle.

Row	Task/Definition	Example Use
3	**Bypass:** A tactical mission task in which the commander directs his unit to maneuver around an obstacle, position, or enemy force to maintain the momentum of the operation while deliberately avoiding combat with an enemy force (ADRP 1-02, 1-11).	

Bypass, by definition, involves deliberately avoiding combat. On orbit, this may involve deliberate maneuvers to steer clear of hostile enemy spacecraft. In the electromagnetic spectrum, a force may choose alternate communications paths to bypass enemy jamming. On the ground, objectives that influence the conduct of space warfare—for example, enemy ground stations or launch infrastructure—may be bypassed by the initial invasion force, only to be occupied and exploited by specialized follow-on forces at a later phase of the operation. Rocket fuel manufacturing and storage depots that contain highly dangerous chemicals are prime candidates for occupation—after the main fight has passed. In this manner, an enemy's space infrastructure may be turned against it in a manner analogous to occupying or seizing an enemy's airfields or port facilities. Indeed, in such a scenario, it is not difficult to imagine special purpose or airborne forces performing this function at the request of Space Command (forces that possibly may be delivered through point-to-point space planes in the future).

Less obviously—but perhaps more importantly during routine competition—bypassing may also include deliberate efforts to avoid observation. On orbit, a communications satellite may seek to avoid passing through a zone in which a signals intelligence satellite or a ground observation post may be capable of observing it. Similarly, a ground station may wish to avoid operating, employ concealment, or regularly change its location to avoid being observed by an imagery collection satellite. Determining when and how an enemy may be capable of observing all segments of space domain systems (ground, link, orbital) is a core function of a space planner and is of critical importance to fighting and winning space wars.

Row	Task/Definition	Example Use
8	**Demonstration:** In military deception, a show of force in an area where a decision is not sought that is made to deceive an adversary. It is similar to a feint but no actual contact with the adversary is intended (ADRP 1-02, 1-27).	

Strictly speaking, the intent of a demonstration is to deceive an adversary. If deception is intended, a demonstration action may draw an adversary's attention away from another activity—perhaps the repositioning of a naval task force. What if, however, the intent of a demonstration on orbit or in the electromagnetic spectrum was not to deceive but either to condition or to message an adversary? Changing the orbital locating of a satellite so that it is nearer to an adversary satellite, for example, may seem suspicious— assuming, of course, that the enemy's space domain awareness capabilities can observe the activity in the first place. These demonstrations of potential satellite attack may draw out an adversary's hidden war-reserve defensive modes on board a satellite, allow friendly forces to determine its reaction times, or simply run down the fuel reserves with its constant maneuvering out of the way of potential threats. This is similar to Cold War tactics of probing enemy air defenses and electronic signatures in peacetime. If such demonstrations become routine, an adversary may gradually become accustomed to them and let down their guard.

In an alternative but equally likely scenario, relocating a satellite near the highly sensitive satellite of an adversary may cause a completely different response: alarm. Such activity certainly invites a risk in escalation, but placing a space-based asset in check may provide a way for belligerents to send a message to an adversary that is not as publicly obvious as moving a carrier strike group. If satellite activity coincides with the movement of a carrier strike group, the adversary's perception may be altogether different, leading to far greater concern and an assumption of a coordinated effort across domains. The danger of misinterpretation becomes very real if space activity *appears* to be synchronized with other-domain activity, implying an altogether different potential for escalation. *Demonstration*, then, seems like the tactical task most nearly able to fulfill an information operations function, but commanders should deliberately desynchronize demonstrations from other potentially provocative activities unless the goal is provocation.

Planners and decision-makers should always consider the space conflict escalation ladder (more on this topic in Chapter 4) when initiating threatening space actions in order to prevent undesired escalation in space and on the earth—even if the actions are seemingly innocuous. As a general rule, certain categories of satellites—those with a missile warning mission, for example—should not be closely approached. The underlying assumption to this rule is that if a nation intended to initiate a nuclear conflict, it would first try to eliminate their enemy's ability to detect missile launches. Thus, under the right circumstances, it is possible that a close approach to a missile warning satellite could be interpreted as the precursor to nuclear war.

Row	Task/Definition	Example Use
9	**Destroy:** A tactical mission task that physically renders an enemy force combat-ineffective until it is reconstituted. Alternatively, to destroy a combat system is to damage it so badly that it cannot perform any function or be restored to a usable condition without being entirely rebuilt (ADRP 1-02, 1-28).	

While an antisatellite missile, a space mine, or a debris field would certainly destroy a satellite, from a space operations perspective, destruction does not necessarily involve blowing a satellite to smithereens. Indeed, as long as no practical way of removing orbital debris exists, such destruction is highly undesirable to space-faring nations. Obliteration aside, however, sufficiently interfering with launch, ground, or orbital operations could very well lead to combat ineffectiveness. Whether through cyberattack or missile strike against a ground station, by frying a satellite's optical sensors via laser (an "attack by fire in order to destroy") or by uploading malicious code to a satellite's on-board computer (or by ensuring the presence of malicious code on a satellite component from the time of manufacture) or by painting a satellite's solar panels or cutting its wires, many options exist to render a space capability useless. The more indirect the option, the more difficult its attribution.

To continue with the theme of "rendering an enemy combat force ineffective," the surreptitious repositioning of satellites provides an interesting option to achieve the intent of destruction (or even disruption). Since the utility of many spacecraft depends upon precise orbits, using a robotic spacecraft or a co-opted ground station to reposition satellites outside their usable orbits renders them ineffective for their intended purposes. The primary benefits of such a course are that the co-opted spacecraft remain usable to the friendly force (like a captured ship), they exist for bartering in conflict negotiations, and they can be returned postconflict if policy aims require it. Just as it may be unwise to leave a defeated enemy without a standing army for self-protection, it may be similarly unwise to leave one without at least some organic space capabilities.

Row	Task/Definition	Example Use
34	**Turn:** A tactical mission task that involves forcing an enemy force from one avenue of approach or mobility corridor to another (ADRP 1-02, 1-94).	

The doctrinal definition of *turn* provides little insight into its true meaning, and in graphical depiction, it is very easy to misinterpret as something like a deep flank attack. In practice, a turning movement involves employing forces against an enemy's lines of communication in such a way as to make their position untenable. The enemy must then alter its plan of attack—a task made easier if it has a robust zone of communications—or else it must withdraw to the best of its ability—a task made difficult because friendly forces are attacking its rearward areas. Generally speaking, a turning movement does not require the overmatch of an envelopment, the intent of which is to encircle and destroy the enemy (if they do not surrender first), but because it places forces in an enemy's rear, it has the potential to disorganize the enemy force, possibly allowing for exploitation.

For space operations, the utility of the turning movement largely depends upon the idea of the data line of communication. In current US doctrine, lines of communication allow forces to bring forward supplies and reinforcements. Not addressed in this definition are the data transmission functions performed by these lines—functions uniquely suited for space and cyber operations through the establishment, maintenance, and protection of data lines of communication. In a multidomain turning movement, a ground force, for example, may attack the enemy main body while space forces attack the enemy's satellite-based data lines of communication. The attack against these data lines of communication may employ numerous tactical tasks—attack by fire, disrupt, destroy—as part of the joint force's larger turning movement.

Fires Support Coordination Measures

As with the maneuver warfighting function, the fires warfighting function comes with numerous control measures. As in the discussion of maneuver, the discussion of fires also begins with a discussion of the control of geographical areas and volumes. In existing military symbology, the airspace coordination area, the free-fire area, the no-fire area, and the restricted area (where some fires are allowed) all have analogues in the application of space operations (Table 25). These areas, like their maneuver counterparts may be rectangular, circular, or irregular in shape depending upon the requirements of the operating environments. They may additionally include such amplifying information as the maximum and minimum altitudes of the area (thereby making it a three-dimensional shape) and the times for which the areas are in effect.

Row 1 of the following table provides an example of a circular space coordination area, which may, given the tendency to measure distances from spacecraft, most lend itself to orbital operations (except for the uniqueness of some orbital motions. See Figure 65, for example, and the discussion on space defense identification zones). The information within the symbol designates it as a space coordination area (SCA) belonging to the First Space Operations Squadron (1SOPS). The relevant altitudes are between 800 and 900 kilometers, and the relevant portion of the electromagnetic spectrum (EMS) is between 1,000 and 1,200 megahertz. The SCA is in effect from January 27 to 28. What is different between this symbol and its counterpart in existing airspace symbology is the inclusion of the portions of the electromagnetic spectrum relevant for the operation.

Row	Description	Notes/Reference	Example Usage
1	Space Coordination Area (SCA)	Adapted from the Airspace Control Area symbol, MIL-STD-2525D, Appendix H—Control Measure Symbols, Table H-XVI, p. 509; may be rectangular, circular, or irregular in shape	SCA 1SOPS MIN ALT: 800 KM MAX ALT: 900 KM EMS: 1000-1250 MHz EFF: 271400JAN21-280600JAN21
2	No Fire Area (NFA)	MIL-STD-2525D, Appendix H—Control Measure Symbols Table H-XVI, p. 514-516; may be rectangular, circular, or irregular in shape	SPACECOM 271400JAN21-280600JAN21
3	Restricted Fire Area (RFA)	MIL-STD-2525D, Appendix H—Control Measure Symbols, Table H-XVI, p. 517-519; may be rectangular, circular, or irregular in shape	RFA CFSCC 271400JAN21-280600JAN21
4	Free Fire Area (FFA)	MIL-STD-2525D, Appendix H—Control Measure Symbols, Table H-XVI, p. 511–513; may be rectangular, circular, or irregular in shape	FFA SPACECOM 271400JAN21-280600JAN21

Row	Description	Notes/Reference	Example Usage
5	Space Electronic Warfare Firing Area (SEWFA)	Adapted from the Position Area for Artillery (PAA) symbol, MIL-STD-2525D, Appendix H—Control Measure Symbols, Table H-XVI, p. 520–521; may be rectangular or circular in shape	SEWFA / SEWFA / SEWFA / SEWFA
6	Fire Support Area (FSA)	MIL-STD-2525D, Appendix H—Control Measure Symbols Table H-XVII, p. 533–535; may be rectangular, circular, or irregular in shape	**FSA MAX** 010700ZJAN08 - 010745ZJAN08
7	Sensor Zone	MIL-STD-2525D, Appendix H—Target Acquisition Control Measure Symbols Table H-XVIII, p. 553–555; may be rectangular, circular, or irregular in shape	020300ZDEC08 - 090500ZDEC08 CENSOR ZONE BRAVO
8	Dead Space Area	MIL-STD-2525D, Appendix H—Target Acquisition Control Measure Symbols Table H-XVIII, p. 550–552; may be rectangular, circular, or irregular in shape	060300ZNOV07 - 090500ZNOV07 DA USEUCOM

Table 25: Fire Areas for Space Warfare. Source: Table compiled by authors with symbols adapted from MIL-STD-2525D, Appendix H.

The inclusion of the electromagnetic spectrum constraints is particularly important for cross-domain electronic warfare activities. Just as an aircraft may be directed to fly through an airspace coordination area to decrease chances of friendly fire, so a spacecraft may be directed to adjust its orbital pattern or its frequency usage to avoid another spacecraft's maneuver or the electronic warfare of a ground unit. In the future, if the DIA's assumptions prove true, a space coordination area could even be used to dictate a safe path for a nearby satellite to avoid the accidental effects of getting too close to an on-orbit jammer or to designate free-passage zones for spacecraft owned by neutral countries.

To further protect transiting satellites—including commercial ones—a space operation may include no-fire areas (NFA, Row 2) or restricted-fire areas (RFA, Row 3). These constraining symbols do not seem necessary for the application of antisatellite missiles, whether orbital or ground based. If the decision to execute such a provocative action is taken, it is presumed that the safety of any nearby orbital assets will be assessed with the decision. In other words, no-fire areas and restricted-fire areas aim to constrain warfare throughout its dynamic execution; antisatellite missiles are not a dynamically employed asset. However, these control measures are likely to be very useful within the realm of electronic warfare. Such areas may exist for a given frequency range for a given time, may be active during certain portions of a friendly constellation's orbital configuration, and may have associated triggers for their discontinuance or reinstitution. In this way, one begins to imagine no-fire and restricted-fire areas cycling on and off as the satellites orbit—blinking control measures. When a new operation or a new phase of operation begins, the laydown of the blinking control measures changes—older areas being deactivated, newer ones being enacted.

Free-fire areas (Row 4) serve as the conceptual opposite of no-fire areas. In a free-fire area, there are no restrictions. In terrestrial operations, any target within the free-fire area may be engaged with any means necessary—cannon, rocket artillery, or close-air support to name a few options. These designations would be according to the currently perceived space conflict level, or even terrestrial conflict level, or the "nearness" of a threatening satellite. In space operations, particularly their electromagnetic aspects, free-fire areas require a distinction. An area may be free-fire for electronic assets but restricted-fire for kinetic assets. Oppositely, an area may be no-fire for UHF jammers, but the existence of such a control measure would be completely inconsequential to a unit operating EHF communications satellites in that sector. This way of thinking presupposes a future in which satellites can be labelled "UHF" or "EHF." What is more likely is that software-defined radios, both on-orbit and on the ground, will become the norm, making communications less susceptible to particular flavors of jammers and making the task of determining what frequency to jam much more difficult. The arms race that follows also will include the development of jammers with software-defined emitters and the proliferation of quantum communications.

The following figure depicts notional restricted-fire and free-fire areas in space. The blue represents multiple NFAs at a given moment in time—generally structured to protect polar assets. Similarly, the orange FFAs exist to allow for the engagement of threats to those polar assets. It is interesting to note that the NFAs and FFAs seem to intersect at this particular moment. This is likely a function of the orbital motions of the satellites and illustrates the need for close coordination in the employment of precision fires. It is also possible that the NFAs are protecting the polar satellites from a particular wavelength of electromagnetic energy or from a particular phenomenology of weapon—types of weapons that may be freely employed against satellites in the orange zone. In this way, NFAs and FFAs may potentially exist in the same physical space.

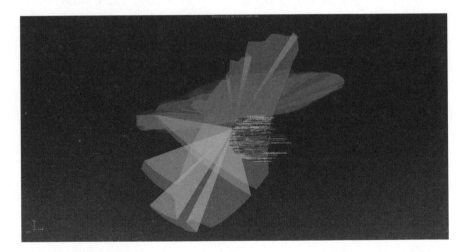

Figure 36: Notional Restricted-Fire (Blue) and Free-Fire (Orange) Areas in Space. *Source:* Graphic created by Paul Szymanski using the Satellite Orbit Analysis Program (SOAP) from The Aerospace Corporation.

Just as it is necessary to determine where a fires asset can and cannot fire, it is also necessary to determine from where the asset will fire. In the employment of artillery, the position area for artillery (PAA) performs this function, proscribing where an artillery unit will establish its firing positions and giving them latitude to reposition their individual cannons within that area. PAAs vary in size depending upon the echelon of the firing unit, and units will often require multiple PAAs to progress them through the operation, moving with friendly forces as they advance or displacing if the threat of counterbattery fire requires. For space operations, the analogue to the artillery's PAA is the space electronic warfare firing area (SEWFA, Row 5). Like the PAA, the SEWFA may be circular or rectangular. For the purposes of optionality, it is worth noting that the naval symbol for fire support area performs the same function. Although traditionally used when ships are firing in support of an amphibious landing, it might be used for space operations if counterspace fires were ever employed from a ship (Row 6).

Of course, the employment of fires assets also requires visibility of the target. Sensor zones (Row 7) depict the area covered by a sensor, such as a ground-based radar. In traditional 3-D visualizations, the sensor zone often appears as a "bubble" for omnidirectional sensors or as a wedge-shape for pointed sensors. Figure 37 and Figure 38 show some space surveillance sensor visibility cones in the western and eastern hemispheres, respectively. Where these sensors cannot see is a dead space area (Row 8), and rather than depicting every single sensor's coverage area, it may be practical to denote only the areas that the sensors cannot cover. It is in these areas that an enemy may maneuver freely, whether on orbit, on the surface, or within the electromagnetic spectrum. Such activities constitute events and require their own symbols.

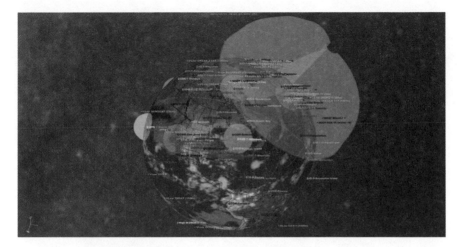

Figure 37: Some Space Surveillance Sensor Visibility Cones (Western Hemisphere). *Source:* Graphic created by Paul Szymanski using the Satellite Orbit Analysis Program (SOAP) from The Aerospace Corporation.

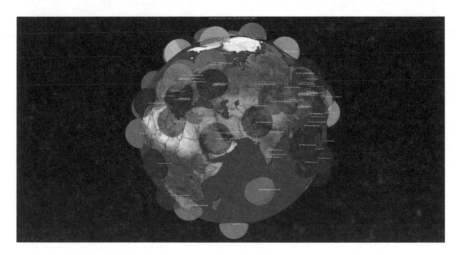

Figure 38: Some Space Surveillance Sensor Visibility Cones (Eastern Hemisphere). *Source:* Graphic created by Paul Szymanski using the Satellite Orbit Analysis Program (SOAP) from The Aerospace Corporation.

Section 5: Event Symbols

Event symbols depict activities of interest and are particularly useful in tracking current events (as opposed to planning future operations). These symbols combine the standard event frames from Table 1 with icons and modifiers to depict events. Unlike standard symbols, which typically—although not exclusively—employ solid black icons, the event symbols proposed here leverage a variety of colors to convey meaning. This characteristic makes them more useful in digital displays, particularly in any system that can automatically cue operators to respond to the event stimulus. From environmental considerations to launches, on-orbit checkout, rendezvous and proximity operations, and changes to weapons release authorization levels, the layout of events in this section follows a general escalation of conflict. It culminates with a discussion of hostile attack events.

Environmental Events

As in other domains, environmental factors remain an early and often-revisited consideration in planning and executing space operations. Generally speaking, environmental factors in space influence operations in a way that is analogous to the way environmental factors on earth influence surface operations, ranging from cancelling a plan to merely inconveniencing it. A hurricane, for example, may delay an operation, but a rainstorm may only cause additional operational friction in military operations. Environmental factors, in other words, may or may not cause a responding action, but they are absolutely essential to achieving situational understanding.

The following symbols represent environmental events that require awareness and possibly action. Rows 1 through 4 depict solar flares of various intensities. These flares are largely unpredictable, and because charged particles travel from the sun to the earth within a matter of minutes, it is probably impractical to do anything to "dodge the bullet." Instead, knowing *that* flares occurred is more useful in postevent analysis. Solar flares, for example, may cause electrical upsets within satellites, either affecting their normal functions or presenting as an enemy's tampering. They may also degrade certain communications frequencies and radar systems. Knowing that solar flares happened helps rule out the likelihood of enemy activity and imprudent response.

More surreptitiously, however, a clever adversary might initiate space attacks under the cover of solar flares to hide his intentions. Depending on the duration and extent of the solar storm, he may even achieve his objectives before the attacked country is able to understand whether it is under intentional attack or what the attackers' goals and motives are (command decapitation; blinding battlefield surveillance/reconnaissance; confusing missile warning; degrading positioning, navigation, and timing services; or delaying weather info, for example).

Row	Symbol	Event
1	TOT: 23 Hrs. 12 Min. 10092900ZMAR29	Solar Flares—Intensity Unknown
2	TOT: 23 Hrs. 12 Min. 10092900ZMAR29	Solar Flares—Intensity Low
3	TOT: 23 Hrs. 12 Min. 10092900ZMAR29	Solar Flares—Intensity Medium
4	TOT: 23 Hrs. 12 Min. 10092900ZMAR29	Solar Flares—Intensity High

Row	Symbol	Event
5		Meteor Showers
6		High Space Debris Zones
7		High Radiation Environment

Table 26: Space Environmental Event Symbols. Source: Created by authors.

In a similar way, the knowledge of meteor showers (Row 5) provides insight into satellite failures. If, for example, all contact with a satellite is lost during a period of high meteor activity, it may be surmised that a collision occurred but was undetected—again casting doubt on the possibility of enemy action or accidental destruction by orbital debris (debris zone icon shown in Row 6). If space-to-earth projectiles ever come into existence—as was proposed in the strategic defense initiative—the knowledge of a meteor shower might minimize the fear that what is coming into the atmosphere are weapons, thereby minimizing the risk of an escalatory response.

A third type of environmental factor that is important to understand is radiation (Row 7). Radiation occurs naturally within the space environment—most notably in the Van Allen radiation belts—and results primarily from solar activity. Understanding normal levels of radiation not only helps in the employment of spacecraft—GPS satellites, for example, are designed to withstand and operate within the radiation environment of MEO—but it provides a baseline for what may be considered abnormal. Abnormal levels of radiation may come from increased solar activity, requiring precautionary measures to protect spacecraft, or in an extreme case, it may come from an exoatmospheric nuclear detonation. Such a detonation would certainly be detected, and mapping its effects on the radiation environment of space would be essential for managing postdetonation response operations and damage assessment effectiveness of subsequent ASAT attacks.

Launch Events

Table 27 depicts major events as they might occur chronologically either in the preparation and launch of a ballistic missile or in the preparation and launch of a multistage space launch vehicle. Row 1 depicts an event symbol for the preparation of an antisatellite missile launch. In any of these cases, the notification of the event would precipitate follow-on actions like battle drills, cueing space surveillance sensors, or the enactment of preplanned response options.

Row	Symbol	Event
1		Ballistic Missile Launch Preparations
2		Ballistic Missile Launch Detected

Row	Symbol	Event
3		Space Launch Preparations
4		Space Launch, First Stage Ignition
5		Space Launch, Second Stage Ignition
6		Space Launch, Third Stage Ignition
7		Antisatellite Missile Launch Preparations

Table 27: Launch-Related Events. Source: Created by authors.

On-Orbit Testing Events

Following the successful launch of a satellite, it begins a period of on-orbit testing before being accepted for operational use. Often, the satellite will conduct a maneuver, possibly consisting of multiple engine burns, to approach and enter its final orbit. Table 28 shows event icons for the initial and final on-orbit tests as well as for the detection of a satellite maneuver. Note that the "Maneuver Detected" icon (Row 3) includes amplifying information about the time of maneuver (TOM), including how long ago the maneuver happened and the actual time (in Zulu time) of the event. A general symbol for "Satellite Status—State Change" is included here (Row 4). This state change may be due to environmental damage, mechanical failure, purposeful reconfiguration of the satellite when initiating alternate modes of operation, unusual maneuvers or attitude changes, or other change in functionality.

Row	Symbol	Event
1	INIT TEST	Initial Orbital Checkout, Currently Testing
2	FINAL TEST	Initial Orbital Checkout, Finished Testing
3	TOM: 9 Hrs. 16 Min. Ago 10091900ZMAR12	Maneuver Detected

Row	Symbol	Event
4		Satellite Status—State Change

Table 28: Symbols for Orbital Testing and Maneuver. Source: Table compiled by authors. The symbol in Row 3 was originally published in Paul Szymanski, "Issues with the Integration of Space and Terrestrial Military Operations," *Wild Blue Yonder Online Journal* (June 2020). https://www.airuniversity.af.edu/Wild-Blue-Yonder/Article-Display/Article/2226268/issues-with-the-integration-of-space-and-terrestrial-military-operations/.

Rendezvous and Proximity Events

Whether for station keeping, to avoid a potential collision, or to change orbital location, generic satellite movements occur frequently. When satellites move in relation to an adversary's satellite, we may more appropriately refer to such movement as a maneuver. When the satellite is deliberately positioning itself to come near and observe another satellite, this is called a rendezvous and proximity operation (RPO). RPOs may happen at greater or lesser ranges, depending on the objectives of the aggressor. As discussed in Section 3, commanders may predefine SDIZs, LREZs, and CAEZs to gauge the level of threat from a particular approaching satellite or to control a friendly RPO. Within a greater space defense identification zone (SDIZ), the volumes of space around a target satellite may be further subdivided into long-range engagement zones (LREZs, greater than 10 kilometers or 20 minutes away) and close-attack engagement zones (CAEZs). The CAEZ is further subdivided into yellow and red categories. In the CAEZ-Yellow, the attacking spacecraft is 10 kilometers or less or 20 minutes or less from its target. In a red CAEZ, the attacker is within one kilometer or five minutes.

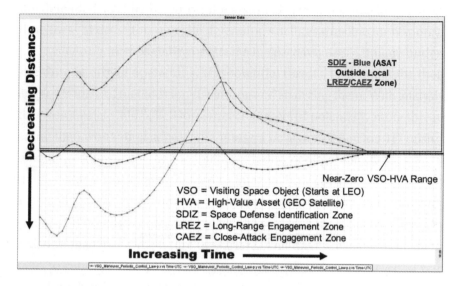

Figure 39: Typical Approach Trajectories. *Source:* Created by Paul Szymanski using the Java Spacecraft Simulation Tool (Jastro) by Barron Associates, Inc.

Figure 39 shows what a typical approach trajectory may look like during an RPO. Each line represents a different coordinate of the spacecraft in three dimensions (x, y, and z) as it transits through SDIZ-Blue toward a high-value asset (HVA). Figure 40 zooms into the final stages of this approach, the thick, black bar representing the target's position. When the satellite crosses CAEZ-Red on the left side of the graph, it is in a close approach, but not all of its coordinates are yet converging on the target. This tendency for RPOs to have a first, close approach is significant for satellite defense; attacking space objects have two chances to attack, one more temporary, the second longer-term. As the satellite transits CAEZ-Yellow, its coordinates converge, meaning that the satellite is gradually nearing a stable position as it enters CAEZ-Red.

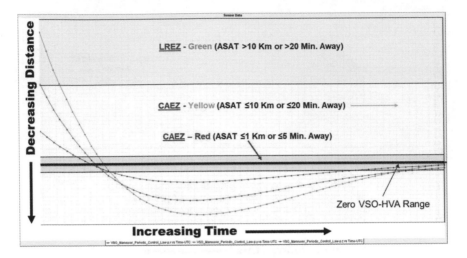

Figure 40: Typical Approach Trajectories—Endgame. *Source:* Created by Paul Szymanski using the Java Spacecraft Simulation Tool (Jastro) by Barron Associates, Inc.

Table 29 shows a series of RPO event symbols. The squiggly arrow is a repurposing of the "foraging/searching" activity icon and suggests the intent of an RPO. True, an RPO could be a prelude to an attack, but it could also be to gather intelligence (such as the detection of war-reserve modes and reactions to close approaches), to telegraph a message to an adversary, to inspire the observed satellite to waste precious fuel reserves maneuvering out of the way, or simply to practice engaging and disengaging a target to make an adversary get used to such maneuvers before a real attack occurs. The symbol in Row 1 accounts for this uncertainty, the white arrow indicating that the type of operation is unknown. In Row 2, the blue arrow indicates that the RPO spacecraft is within the designated SDIZ but is not within a designated LREZ or CAEZ. The colors change as the approaching craft gets nearer and nearer to its target, with the yellow and red arrows corresponding to CAEZ-Yellow and CAEZ-Red.

Row	Symbol	Event
1	TOT: *9 Hrs. 16 Min.* *10094900ZMAR23*	RPO, Type Unknown
2	TOT: *9 Hrs. 16 Min.* *10094900ZMAR23*	RPO within Designated SDIZ
3	TOT: *9 Hrs. 16 Min.* *10094900ZMAR23*	RPO within Designated LREZ (> 10 Km or > 20 Minutes Away)
4	TOT: *0 Hrs. 19 Min.* *02154400ZFEB14*	RPO within Designated CAEZ- Yellow (≤ 10 Km or ≤ 20 Minutes Away
5	TOT: *0 Hrs. 4 Min.* *02154400ZFEB14*	RPO within Designated CAEZ Red (≤ 1 Km or ≤ 5 Minutes Away)

Table 29: RPO Events. Source: Created by authors.

It is important to note here that while deliberate RPOs may not necessarily be numerous, encounters among satellites are exceedingly common. If a commander desired to receive a notification whenever one space object was within 5 km of another space object, for example, he would receive several thousand reports per month. Figure 41 illustrates the frequency of close encounters between space objects due to simple natural orbital motions employing real data. The bottom axis shows the distances between satellites in the encounters (between 200 and 4,900 meters), and the left axis shows the number of space objects within that range bin.* As one may expect, as the range increases, more satellite encounters are experienced within that volume of space. For example, there were roughly ten times during this particular month when satellites came within 300 meters of each other. At the opposite end of the graph (right), there were more than 350 encounters between 4,800 and 4,900 meters and almost 350 more encounters between 4,900 and 5,000 meters.

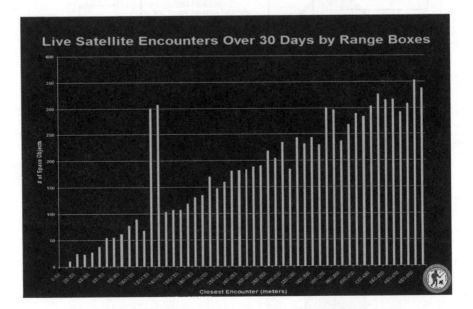

Figure 41: Past Close Approach Statistics due to Natural Orbital Movements. *Source:* Graphic created by Paul Szymanski using the Satellite Orbit Analysis Program (SOAP) from The Aerospace Corporation, orbital data from https://www.space-track.org, and the Space Warfare Analysis Tool (SWAT) developed by Paul Szymanski.

---

*   This is a product of the Space Warfare Analysis Tools software created by Paul Szymanski.

Weapons Release Authorization Events

As tensions escalate on orbit, it will be essential for operators to understand what permissions they have for engaging other satellites, especially if future satellites are equipped with an automated shoot-back capability. The event icons to account for such activity are admittedly forward-looking, but based on the Defense Intelligence Agency report, it seems more prudent to establish such a system and not need it than to need it and not have it in place. Table 30 shows event symbols for orbital attack operations. The example symbols include a red (enemy) ASAT icon within the event frame, highlighting the fact that there has been a change to its weapon release authorization level (positive control, autonomous operation, weapons hold, weapons tight, or weapons free). Although discussed more in-depth in subsequent chapters, the definitions of these terms are included here so that the reader may more readily make sense of the symbols' meanings. The definitions themselves derive from Joint Publication 3-01.1, *Aerospace Defense of North America*, and have been modified for space-specific use.

Space Autonomous Operation—In space defense, the mode of operation assumed by a space system after it has lost all communications with human controllers. The space system assumes full responsibility for control of weapons and engagement of hostile targets in accordance with on-board surveillance and weapon system control logic. This automatic state may occur on a regular basis because of orbital movements outside regions of ground coverage and control.

Space Positive Control—A method of space control that relies on positive identification, tracking, and situation assessment of spacecraft within a space defense area, conducted with electronic means by an agency having the authority and responsibility therein.

Space Weapons Hold—In space defense, a weapon control order imposing a status whereby weapons systems may only be fired in self-defense or in response to a formal order.

Space Weapons Tight—In space defense, a weapon control order imposing a status whereby weapons systems may be fired only at targets recognized as hostile.

Space Weapons Free—In space defense, a weapon control order imposing a status whereby weapons systems may be fired at any target in orbital space of defined altitude and inclination, not positively recognized as friendly.

A banner across the top of the domain frame denotes the specific change in weapons release authorization. In these example symbols, the geometric shape specifying the rule of engagement authorization level sits atop the banner for illustrative purposes (recall that the ROE amplifiers were depicted in Table 15). In general, weapons release authorization levels will become freer as ROEs become less restrictive, but this may not always be the case. It would be theoretically possible, for example, for a disruption weapon (a fairly low-level attack) to be authorized a weapons' free status. In other words, one level of ROE does not always correspond to the same weapons release authorization level. Although situations will undoubtedly vary, weapons release authorizations should roughly correlate to the currently perceived threat level and the corresponding rung on the conflict escalation ladder for both space and terrestrial theaters.

Row	Symbol	Event
1	△ **Positive Control** **ASAT EW**	Authorization Level Change, Positive Control
2	▲ **Autonomous Operation** **ASAT EW**	Authorization Level Change, Autonomous Operation

Row	Symbol	Event
3	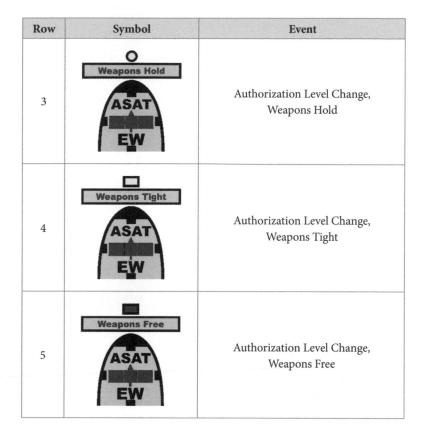	Authorization Level Change, Weapons Hold
4		Authorization Level Change, Weapons Tight
5		Authorization Level Change, Weapons Free

Table 30: Example Symbols for Weapons Release Authorization Levels. Source: Created by authors.

Hostile Attack Events

With the loosening of weapons release authorization levels, hostile attack events become imminent. How then might we account symbolically for actual attacks by antisatellite weapons? Table 31 depicts such symbols beginning with two events that may be the first indicators of an antisatellite attack: the tracking, telemetry, and control (TT&C) commands to initiate an aggressive maneuver or to execute an attack (Rows 1 and 2). The utility of these symbols presupposes the ability to collect such signals and the knowledge of what the intercepted command signal means. Such should be the intelligence goals of space operations forces.

For subsequent rows, the attack iconography can work for either the aggressor or the aggressed. Rows 3 and 4, for example, depict enemy satellite events—kinetic kill vehicle and grappler attacks, respectively—which could either suggest they are the ones doing the attacking or that they are the one being attacked. Alternatively, the symbols in Rows 5 and 6 depict laser attacks (low- and high-power, respectively) on an optical imager, suggesting that the blue satellite is the victim.

Rows 7–10 show other types of electronic activity events, including the painting of a satellite by radar (Row 7), an attack by a jammer or cyber actor (Row 8), and a high-power microwave attack (Row 9). Row 10 shows an event symbol for a detected change in a satellite's electronic signature, possibly an indicator that it is engaging in an imminent attack.

Finally, Row 11 accounts for the possibility that the type of ASAT being employed is unknown.

Row	Symbol	Event
1		ASAT TT&C, Command Maneuver
2		ASAT TT&C, Command Attack
3		ASAT Attack, Kinetic Kill Vehicle

Row	Symbol	Event
4		ASAT Attack, Robotic Grappler
5		ASAT Attack, Low-Power Laser
6		ASAT Attack, High-Power Laser
7		RADAR Paint
8		Jammer/Cyberattack

Row	Symbol	Event
9	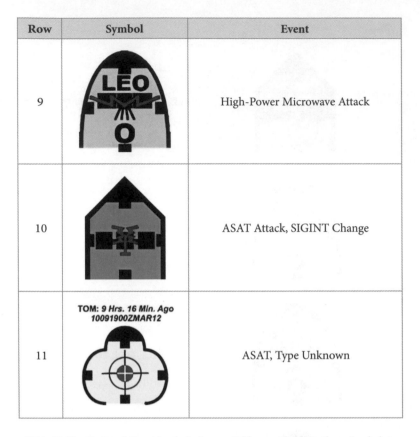	High-Power Microwave Attack
10		ASAT Attack, SIGINT Change
11		ASAT, Type Unknown

Table 31: Hostile Attack Event Symbols. Source: Table compiled by authors. Symbols in Rows 3, 4, and 10 originally published in Paul Szymanski, "Issues with the Integration of Space and Terrestrial Military Operations," *Wild Blue Yonder Online Journal* (June 2020). https://www.airuniversity.af.edu/Wild-Blue-Yonder/Article-Display/Article/2226268/ issues-with-the-integration-of-space-and-terrestrial-military-operations/.

Hostile attack events on orbit necessarily require significant analysis, linking the particular action to the objectives they support. Figure 42 and Figure 43 provide insight into the complexity of this weaponeering analysis. The following screenshots from the Space Warfare Analysis Tools (SWAT) software developed by Paul Szymanski account for optimizing—among other things—orbital parameters, the statuses of the satellites components, on-board fuel and power reserves, maneuver burn times, available weapons systems, cost-exchange ratios, attack start- and end-time optimizations, attack duration, type of kill desired, and coverages of adversary space surveillance sensors of orbital attacks directed against their own satellites. These activities, of course, must be synchronized in time, space, and purpose with other orbital activities and with the other activities of the joint force.

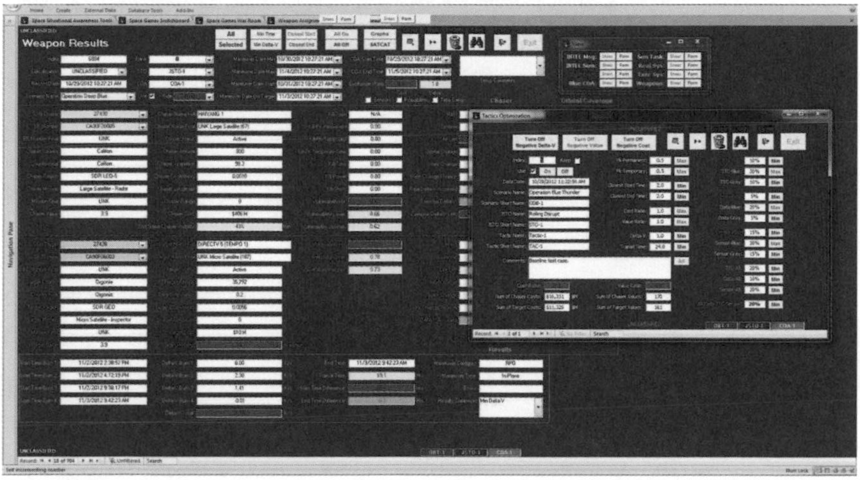

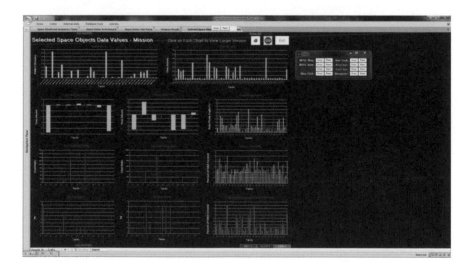

Figure 42: Notional ASAT Attack Optimization Parameters. *Source:* Graphic created by Paul Szymanski using the Space Warfare Analysis Tool (SWAT) that he developed.

Figure 43: Notional ASAT Attack Optimization Statistics. *Source:* Graphic created by Paul Szymanski using the Space Warfare Analysis Tool (SWAT) that he developed.

Conclusion

In any domain, the tactical level of warfare involves the actions of individual service members, of pieces of equipment or craft, and even of units composed of tens of thousands of individuals with hundreds of thousands of pieces of equipment. The tactical level's vastness is bounded by the types of objectives it achieves directly or aids in achieving directly: tactical objectives and operational-level objectives.

As depicted in the schema at the beginning of the chapter, the tactical level of warfare for space operations requires the development of unique courses of action that involve tactical activities like the movement of individual spacecraft, the employment of surface units, and maneuver across the electromagnetic spectrum. In a word, space operations, like all other military operations, boil down to a series of fire and maneuver problems on orbit, on the earth's surface, and within the electromagnetic spectrum—and in the future, areas around the moon. Planning and executing these actions require a symbolic lexicon that builds on existing symbology, expanding and creating new vocabulary that, with some training, is intelligible across the entirety of the joint force.

As we are a species that depends upon encoding information visually, the symbols explored in this chapter—and the many symbols for other-domain operations that were not herein explored—provide an important component to visualizing and understanding the possibilities of space warfare. This symbol set is all the more critical because even experienced warfighters with a keen grasp of tellurian warfare have not directly sensed the peculiarities of orbit or the complexity of the electromagnetic spectrum. Spacecraft and unit symbols represent the means of space warfare that must be employed within a tactical context if any objective is to be achieved.

The control measures—whether for divvying up the space area of operations, defining the paths to be taken, or dictating specific activities—orchestrate the ways in which the means are employed to achieve tactical objectives. The success or failure of the tactical actions requires active tracking throughout the operation and ongoing assessments based on predetermined indicators. Event icons and amplifying information aid in these functions.

If successful, a new or follow-on operation begins—one that is probably already in the planning stages. If unsuccessful, it's back to the drawing board.

Importantly, these tactical actions do not achieve strategic objectives in and of themselves. Even in an operation that involves the close approach to a satellite of national importance, the tactical act of maneuvering the satellite does not achieve a strategic objective independently. Such achievements come by combining tactical actions—often of less visible activities like tasking sensors, optimizing communications, enhancing force protection measures, and employing deception activities. Each of these operations involves tactical actions, which, when combined with sufficient breadth, may contribute indirectly to the achievement of strategic objectives. This more comprehensive orchestration is the province of the operational level of warfare and operational art, the subjects of the next chapter.

Vignette: Has China Prepositioned to Attack Global Positioning System Satellites?

For the United States, the loss of space capabilities would significantly impact its ability to conduct terrestrial warfare. Without global communications, the ability to precisely track its forces, or the intelligence that derives from satellites, the joint force—to say nothing of the rest of the government—would struggle to gain and maintain the initiative and impose its will over its enemies. At best, the additional operational friction would cause increased casualties, delayed results, and undesirable political fallout. At worst, such a loss would cripple the technologically dependent force, making it incapable of winning the war.

Even though the simultaneous loss of all space-based capabilities seems unlikely, the loss of key systems at precisely the right moment may be sufficient to achieve strategic shock. If orchestrated with surprise and coordinated with terrestrial attacks, the effects are likely to be even more devastating. It is the fundamental goal, therefore, of military strategy to prevent adversaries from achieving such surprise. It is the secondary goal of military strategy to seek positions that put friendly forces in a position of advantage vis-à-vis those adversaries. An open-source analysis of Chinese satellite maneuvers in 2011 detected anomalous behavior, suggesting that Chinese military strategists were pursuing such strategic goals.

When one views the orbits of Chinese BeiDou satellites and Global Positioning System (GPS) satellites in 2011, it does not at first appear that there would be any resource conflict. The constellations, though each containing dozens of satellites, did not intersect. They appeared to occupy distinct portions of outer space, each providing independent positioning, navigation, and timing information to their respective countries and to other authorized users (Figure 44).

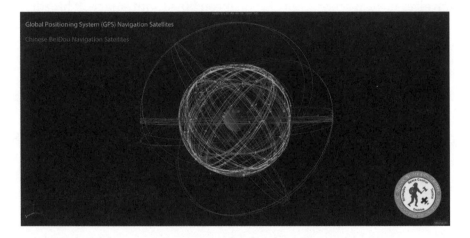

Figure 44: GPS and BeiDou Navigation Satellite Constellations. *Source:* Created by Analytical Graphics, Inc. using their Systems Tool Kit software.

Upon closer examination, however, a different interpretation emerges. Over a thirty-day period, every GPS satellite comes within 250 to 2,500 km of a BeiDou satellite (see Figure 45). Although these distances sound fairly large, because of the orbital patterns of the satellites, a relatively small amount of fuel would be necessary to maneuver each BeiDou to intercept an American counterpart. Without the proper space domain awareness infrastructure and practices in place, the decision of an adversary to make the small adjustment and set its satellite on a collision course—or at least drop off covert space mines in the vicinity of a target—with another one could remain undetected. Known gaps in sensor coverage, the confusion of space debris, and the challenge of identifying objects that may move unpredictably or change orientation compound the problem. Indeed, there are many objects—spent booster stages, for example—that are not regularly tracked but could provide camouflage for attack vehicles.

The occasional, unintentional collision of satellites—such as COSMOS 2251 and Iridium 33 in 2009—provide examples of what an intentional attack could look like—not sneaking up on a target and choosing the proper moment to attack but setting a course and allowing gravity to take the spacecraft "full steam" ahead. This tactic, of course, assumes that the Chinese would be willing to sacrifice one of their own satellites to accomplish the destruction of another—or perhaps multiple others if the debris field spreads. In the latter case, the loss of one satellite may be justifiable. After all, one need not destroy the entire constellation to achieve an advantage, particularly if that advantage corresponds to a window of opportunity during a Pacific Rim battle.

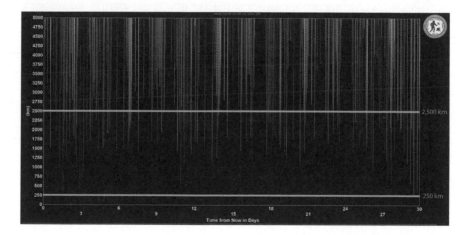

Figure 45: Over Thirty Days, BeiDou Satellites Approach Close to Each of the GPS Satellites. *Source:* Graphic created by Paul Szymanski using the Satellite Orbit Analysis Program (SOAP) from The Aerospace Corporation and orbital data from https://www.space-track.org.

But which satellite to sacrifice? It seems that one would not likely give up a satellite vital to the function of the overall system. Perhaps an orbital spare could fulfill the purpose or, if combined with a deception operation, a zombie satellite. Maybe additional, specialized BeiDou satellites were placed in orbit for the specific purpose of eliminating GPS satellites at a strategic point in time. One such satellite, BeiDou 1D could potentially have been a zombie—one that was apparently dead but could come back to life. Originally in geostationary orbit, BeiDou 1D entered into a graveyard orbit (i.e., one for parking satellites that are no longer useful) much earlier than anticipated based on the engineering design life of the spacecraft.

When a satellite enters a graveyard orbit, it is safe from accidental interference with operational spacecraft, and control of it is typically abandoned. Over time, due to gravitational perturbations (primarily of the earth, sun, and moon) and due to solar radiation pressure, the orbital parameters of the spacecraft become more irregular. In other words, the satellite will drift away from its initial path, and its orbit will become less and less circular. Figure 46, however, shows that, even though BeiDou 1D was listed as "failed," its inclination and eccentricity were not degrading. In fact, they were becoming more regular, which suggests that, despite its failed status, it was still being controlled. A satellite that can be controlled still has utility. Perhaps it was hiding among other satellites and space debris, adopting a temporary but unusual orbit to posture for a surprise attack, or perhaps it was simply floating alongside another piece of debris waiting for natural perturbations to bring it nearer to its target.

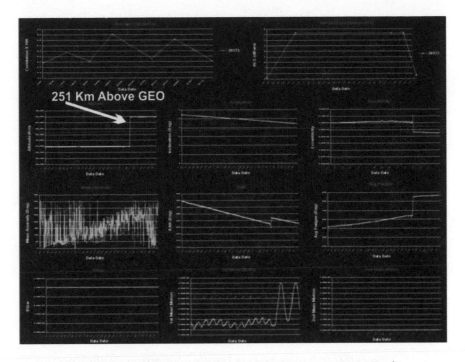

Figure 46: Historical Analysis of BeiDou 1D Orbital Parameters. *Source:* Graphic
created by Paul Szymanski using the Space Warfare Analysis Tool (SWAT)
that he developed and orbital data from https://www.space-track.org.

What if BeiDou 1D was the one that was observed, but there were other
satellites designed specifically to *not* be seen, avoiding sensor detection,
maneuvering at the earth's poles or in the southern hemisphere where
surveillance assets are scarce? Due to the openness of space, it sounds
reasonable that detection and tracking of a satellite should be simple. For
good or ill, however, this is not the case. Any kind of maneuver can confuse
the next set of sensors that may not be able to correlate what it is seeing
with the data of the previous sensor, and even small, continuous thrusts
over time may be enough to change the satellite's paths so that the orbital
equations used to predict future positions are no longer valid. The GOCE
earth resources satellite actually used continuous thrusting for almost five
years to complete its mission, so such tactics are certainly possible, but even
without such tactics, it is very easy to lose track of orbital objects.

Figure 47 depicts the number of lost space objects for a given set of altitudes (light-blue bars), and for the same altitudes, the average radar cross

sections (RCS) of the objects at those altitudes (purple bars). For a single satellite, a relatively large radar cross section (in decibels per square meter, dBsm) means that the object will reflect a significant amount of radar energy, thereby making it more easily detected. The one satellite between 100 and 200 km illustrates this case (left side of the graph). With an RCS of 2.5 dBsm, it should be easily detected, but it was nonetheless lost. As another example, between 20,000 and 25,000 km, there existed eleven lost satellites with an average cross section of 2.1 dBsm. Because this is an average, it is possible that some of those eleven have relatively small cross sections—making them less easily detected—with the larger ones raising the average. Regardless, the point remains that several orbital altitudes contain lost satellites whose average radar cross sections should make them easily detectable, yet they remain lost because of unique orbital characteristics or limitations of the sensor network.

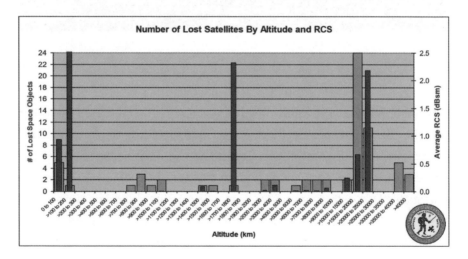

Figure 47: Satellite Orbital Catalog Missing Space Objects. *Source:* Graphic created by Paul Szymanski using the Satellite Orbit Analysis Program (SOAP) from The Aerospace Corporation and orbital data from https://www.space-track.org.

Because the potential to remain lost is great even under the best circumstances, an adversary could, through its knowledge of the Space Surveillance Network (SSN), remain hidden for years. The SSN has known, fixed sites with incomplete global coverage that must contend with tens of thousands of space debris objects and try to determine what is operational, what is dead, and what may constitute a threat. Simply orienting the attitude of a satellite's flat surfaces directly away from an adversary's surveillance systems can significantly reduce its ability to be detected (down to the eighteenth order of magnitude in optical brightness), without any kind of radar or optical absorbing materials.* In addition, satellites are getting smaller and smaller (actually hearkening back to the small size of satellites launched at the dawn of the space age, though, with vastly improved technologies now). These will prove very difficult to detect when tens of thousands of miles away (slant range) from earth-based sensor networks. This difficulty is especially great at geosynchronous orbits where most of the world's communications satellites reside and even more so beyond geosynchronous toward the moon. Given the complexities of the numbers, types, materials, spins, structural configurations, reflective properties, and other characteristics of space objects, the goals of space domain awareness are difficult enough, even without a thinking enemy actively avoiding detection.

What's more, there is no end to the creativity an adversary could employ in disguising its orbital activity, and such attempts at on-orbit systems are certainly not new. Figure 48 illustrates a 1970s patent for a possible stealth satellite system that would try to minimize its radar and optical signatures.† Even dedicating a small portion of a nation's space engineers and scientists to developing covert space weapons could lead to significant gains over time. With such dedication, it would not be overly difficult to develop an orbital system that could fool a night-watch captain at a space operations center into thinking that everything is nominal, even though he is about to witness the destruction of billions of dollars' worth of space assets. Technology, of course, is not a silver bullet. Experience, doctrine, and sheer numbers all play a factor—as they have done throughout the history of warfare in other domains. The United States may have superior technology in space,

---

*   For a light-absorbing carbon nanotubes article, see Lori Keesey and Ed Campion, "NASA Develops Super-Black Material that Absorbs Light across Multiple Wavelength Bands," NASA, Goddard Release No. 11-070, August 11, 2011, https://www.nasa.gov/topics/technology/features/super-black-material.html.

†   Leonard David, "Anatomy of a Spy Satellite," Space.com, January 3, 2005, http://www.space.com/637-anatomy-spy-satellite.html.

and more satellites, but without adequate doctrine and training to guide its decision-making processes in reacting to space warfare, there is a good chance for spectacular failure.

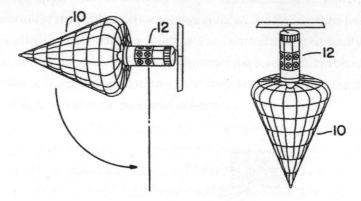

Figure 48: US Patent Schematic for Possible Stealth Satellite. *Source:* Leonard David, "Anatomy of a Spy Satellite," Space.com, January 3, 2005, http://www.space.com/637-anatomy-spy-satellite.html

An analysis of historical satellite catalog data from the North American Aerospace Defense Command (NORAD) demonstrates how, within a given volume of space, numerous potential sources of attack may exist. A traditional computer display depicting a satellite attack may look something like that shown in Figure 49, which is not at all helpful in visualizing military operations on orbit. Several better ways exist that can depict what is happening on orbit in a more meaningful way.

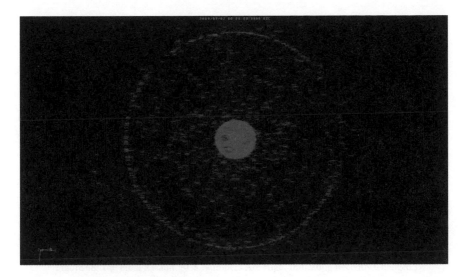

Figure 49: Traditional Orbital View of Notional Satellite Attack. *Source:* Graphic created by Paul Szymanski using the Satellite Orbit Analysis Program (SOAP) from The Aerospace Corporation and orbital data from https://www.space-track.org.

The altitude/inclination plot in Figure 50 shows a volume of space between 600 and 900 kilometers and includes satellites and space junk with inclinations between 98 and 99 degrees—as they existed in 2006. All of the hundreds of objects depicted within this maneuver envelope (outlined by a green oval) could reach all other objects with relatively little change in velocity (delta-v)— in this case, with 300 feet/second of delta-v. Each object, even if not an attack vehicle, poses a potential threat, if only through accidental collision.

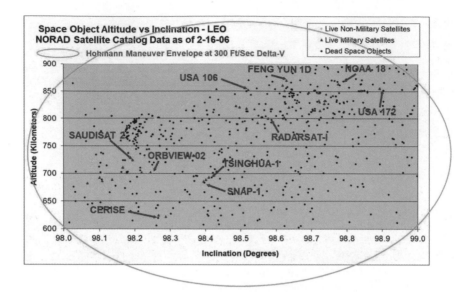

Figure 50: All Objects within the Defined Maneuver Envelope Are Potential Threats.
Source: Graphic created by Paul Szymanski using the Space Warfare Analysis Tool
(SWAT) that he developed and orbital data from https://www.space-track.org.

Because of the density of satellites in sun-synchronous orbit, one satellite playing dead could easily reach and attack many of these critical satellites before ground space surveillances assets can detect and react to this attack. If the attack commences over the South Pacific or Antarctic portion of its orbit where there are no space surveillance assets, the attacker will have the element of surprise, greatly increasing the distance he is able to cover before possible detection by an SDA network. Figure 51 shows this same bunching of satellites for sun-synchronous polar orbits and the amount of fuel (delta-v) that a threat satellite would use to get to the various imagery satellite platforms illustrated in this orbital graphic. Any satellite inspector in the local area might have a difficult time understanding whose satellite is whose, especially on short timescale when various threat satellites and decoys have recently maneuvered (that is, since the last time they were observed and their orbital elements calculated). Even under the best conditions, there is no such thing as "complete" space domain awareness. In one instance, a US Air Force officer recounted a story about the Soviet MIR space station discarding the "honey bucket" from its human waste disposal system. NORAD tracked this object thinking that it was the very large space station until someone noticed the radar cross section was not large enough to be an actual space station, and NORAD fixed its mistake.

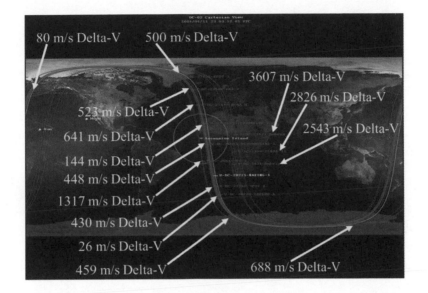

Figure 51: Δv Needed for Hohmann Transfer Rendezvous at Sun-Synchronous Orbit. *Source:* Graphic created by Paul Szymanski using the Satellite Orbit Analysis Program (SOAP) from The Aerospace Corporation and orbital data from https://www.space-track.org.

While these previous two visualizations are certainly helpful (much more so than the nondescript purple dots of Figure 49), the most useful method for visualizing a satellite attack is the one shown in the SAW (Satellite Attack Warning) screenshot of Figure 52. SAW employs the graphical notation explained in the previous chapter on an altitude-inclination plot that shows a space choke point. In this case, the common use of these low altitudes (allowing for imagery collection) and high inclinations (that allow them to revisit the same locations on earth under similar lighting conditions) make for a very popular spot in space. As this depiction makes clear, this volume of space contains a number of friendly, enemy, neutral, and unknown spacecraft, including functional craft and a large number of trackable pieces of orbital debris. The squiggly lines show the observed path of drift for each satellite.

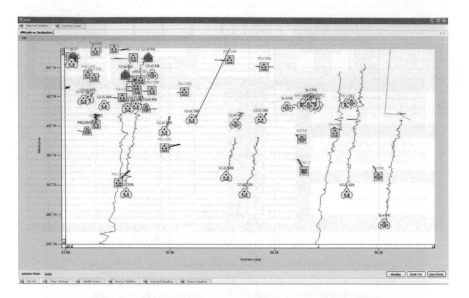

Figure 52: SAW Satellite Choke Point Map—Notional Situational Example. *Source:* Graphic created by Paul Szymanski using the Space Warfare Analysis Tool (SWAT) that he developed and orbital data from https://www.space-track.org.

In *Challenges to Security in Space*, the DIA suggests that orbital threats like kinetic kill vehicles (KKV)—a detachable payload designed to seek and destroy an enemy satellite—may already be in development.* What if an adversary inflated a decoy from a threat satellite while over Antarctica (no terrestrial space surveillance assets cover this region) and then ejected a much smaller space mine to do some future mischief? Would the United States detect this ruse and take appropriate actions? At the same time, many rocket boosters have multiple satellite deployments. Can we be sure that we have observed all of them? As mentioned previously, simulation data suggests that a large-scale orbital attack can begin and conclude within a twenty-four-hour period. With such a capability, it would not even be necessary to destroy one's own satellite in an attack; it would only be necessary to maneuver close enough, employ the KKV, and continue about the original

---

*   Defense Intelligence Agency, Challenges to Security in Space, (Washington DC, 2019), 10, https://www.dia.mil/Portals/27/Documents/News/Military%20Power%20Publications/Space_Th reat_V14_020119_sm.pdf.

mission. What if a zombie satellite like BeiDou 1D was not a positioning, navigation, and timing satellite at all but a decoy host for a KKV? What if such a satellite—or even multiple satellites—are already on orbit, but they have remained undetected, acting like space debris or spent rocket stages?

A swift and undetectable attack against a space-dependent country could cause great disruption—if not operational or strategic shock—and provide the attacking party a distinct advantage for a critical period of time. Attacking as swiftly and as undetectably as possible will certainly contribute to the advantage, and although it is difficult to quantify the advantage gained by a space attack, one Defense University study of major conflicts of the twentieth century showed that, on average, surprise attacks gave a favorable 1:14.5 casualty ratio compared to a 1:1.7 casualty ratio for attacks without surprise. An attack in space may precede terrestrial attacks to blind an opponent, deny or delay communications to forces, or deny indications and warnings of a strategic attack. A strong response to space aggression could potentially deter any terrestrial escalation—an example of immediate escalation to ultimately deescalate a conflict.

With a little imagination, we might invent a few more techniques for the employment of antisatellite weapons, radar avoidance techniques, or techniques for the employment of satellites in general. Here are a few.

1. **Double the Trouble:** A direct-ascent antisatellite missile carries two KKVs. The second KKV holds back in case the first fails. If the first one succeeds, the second hides in a debris cloud for a few days and then maneuvers away when out of sensor coverage.

2. **Hide-and-Seek:** A microsatellite covertly deploys from a multipayload booster and then attaches itself to an old rocket booster—possibly one as benign and uninteresting as an exhaust cone from a 1960s launch—and drifts with it for years. When an opportunity arises, it departs its host outside of sensor coverage and maneuvers toward its target. Since these exhaust cones are as big as a room and are part of a booster that has already completed half of the Hohmann transfer maneuver to a possible target orbit, they would make ideal hiding

spots. Furthermore, no one is likely to waste the maneuver fuel necessary to inspect all of these boosters up close, especially when they are in orbits poorly matched to any potential satellite inspector vehicle.

With the millions of pieces of space debris, an adversary can easily hide among this space junk. Some pieces of debris—delaminated thermal insulation sheets, for example—are heavily influenced by solar pressure and do not follow classical Keplerian orbital dynamics exactly, which will confuse orbital predicational algorithms. Combining the "Double the Trouble" and "Hide-and-Seek" concepts, it would be possible for one booster to hold two ASAT seekers so that the second one can hide in the debris field created by the first attack. Given the "closeness" in altitude-inclination space of some military satellites to many other supposedly "dead" satellites, space junk, rocket boosters, and live satellites from other countries, it would be easy to hide in this Sargasso Sea of dead space objects while waiting for the optimal attack moment. It would be a very difficult problem to figure out which orbital objects are benign and which are threats, especially during high-paced orbital attacks that most likely will occur when out of view of space surveillance assets. In addition, a potential threat satellite will not have a red star painted on its side for easy identification like some MIG fighter aircraft. As a matter of fact, probably a good percentage of the components on a threat satellite would come from friendly Western manufacturing sources.

3. **Shot out of the Blue:** The region of space between geosynchronous orbit and the moon is not routinely monitored. At very high altitudes, very little change in velocity is required to make large changes in inclination or altitude. By taking advantage of its orbital motion, an unobserved antisatellite weapon could make a surprise attack on an asset in the GEO belt. Although there are no indications that any nation is preparing for a "glancing" attack against a GEO asset, an ASAT could dramatically accelerate from a position in translunar space and attack GEO targets with very little delta-v. In fact, because of the centrality of gravity to orbital motion, it actually takes more

fuel to get to GEO than to orbit the moon. In fact, in 1998, a crippled satellite was actually slung around the moon twice to achieve geosynchronous orbit.* Similarly, it takes less delta-v to change inclination the higher in altitude a satellite is. Placing ASATs at Lagrange points, therefore, would put them at the top of the energy peak, and it would take very little to push them back down the energy well toward GEO. Unconventional orbital locations in the vastness of translunar space—at Lagrange points or within some highly eccentric orbits—would make it difficult for space domain awareness assets to observe and calculate an accurate orbital dynamics algorithm solution, thus making them difficult to track.

4. **Equatorial Cutter:** The region of space at low inclinations (at small angles relative to the equator) and low altitudes are not routinely monitored. All low-earth orbit satellites, however, pass through this region. An ASAT here could have access to many LEO satellites but would not be easily tracked.

5. **Space Flyer:** As mentioned in the discussion of the GOCE earth resources satellite, constant thrusting could confuse tracking efforts, making it difficult to determine what the object is, who owns it, where it came from, and where it is going.

6. **Pixie Dust:** One carrier satellite ejects electrostatically charged pieces of fiber optic strands toward a target's on-board star sensors. Such a technique may be enough to confuse the target's attitude control system, altering its orientation or even causing it to tumble.

7. **Hopper:** This is a technique specifically to avoid future detection by a ground-based radar. Figure 53 illustrates the technique. In Frame 1, the satellite passes through a green sensor bubble. In Frame 2, the

---

*   Satbytes, "Earth Satellite First Orbits Moon Twice," htttp://www.spacetoday.org/Satellites/SatBytes/MoonCommsat.html.

satellite's thrusters fire such that by the time it once again breaks the sensor bubble, it is at a significantly different altitude. After another orbit, the satellite is altogether outside of the sensor's range.

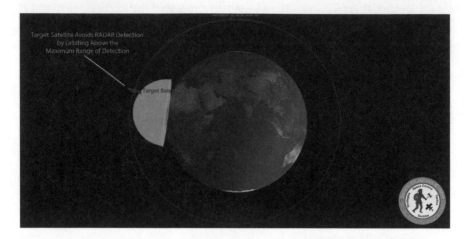

Figure 53: The Hopper Technique to Avoid or Confuse Sensors. *Source:* Graphic created by Paul Szymanski using the Satellite Orbit Analysis Program (SOAP) from The Aerospace Corporation and orbital data from https://www.space-track.org.

8. **Slider:** This technique takes advantages in radar fan patterns. The left frame of Figure 54 shows a sensor cone that extends downward 90 degrees from vertical (tangential to the surface of the earth). With a change in the radar fan to an 80-degree sensor cone (right frame), greater potential exists for a satellite to slide through the gap between the surface of the earth and the lower edge of the beam pattern.

Figure 54: Slider Technique to Avoid Sensors. *Source:* Graphic created by Paul Szymanski using the Satellite Orbit Analysis Program (SOAP) from The Aerospace Corporation and orbital data from https://www.space-track.org.

9. **Skipper:** The skipper technique is similar to the hopper, but instead of shifting vertically, the satellite shifts laterally through the sensor's beam pattern. Figure 55 shows this progression. In the first frame, the satellite ground track (its path) touches the eastern edge of the sensor's beam pattern (the circle). In the second frame, the satellite has naturally drifted into the sensor's coverage. The sensor detects it and predicts the next expected sensor hit. In Frame 3, however, the satellite executes an impulse burn over Antarctica. By the time it reaches the sensor coverage, it is significantly further west than the sensor network's algorithms predicted. By Frame 4, the satellite has exited the radar fan and is not trackable by that radar.

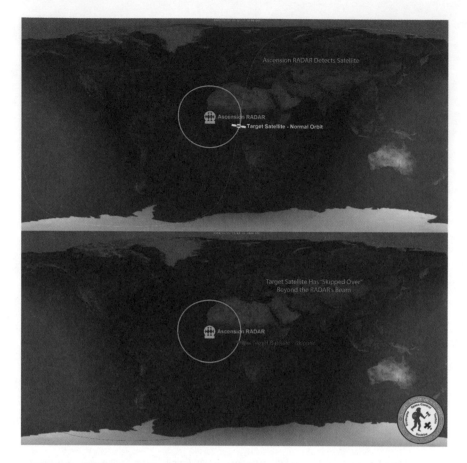

Figure 55: The Skipper Technique Moves the Satellite Laterally Avoiding a Radar Fan. *Source: Graphic created by Paul Szymanski using the Satellite Orbit Analysis Program (SOAP) from The Aerospace Corporation and orbital data from https://www.space-track.org.*

Tracking these kinds of activities and interpreting them requires new sensors, new algorithms, and new visualization techniques. The two following figures show two different ways in which one might monitor and assess the activities of neighboring space objects. Figure 56 depicts a threat assessment chart, with four quadrants: live, dead, debris, and boosters. The friendly asset requiring protection is in the center of the chart with a handful of nearby objects within the close attack engagement zones (either red or yellow) and the long-range engagement zone (green). The orientation of the arrows, their colors, and their amplifying data provide information about the current scheme of maneuver and allow for an interpretation of an attacker's intent. In fact, this particular array shows a live enemy satellite, a nearby enemy satellite playing dead, and an enemy booster—all with ranges closing on the friendly asset. Using many such charts to visualize the threats to multiple assets may give insight into the enemy's operational-level scheme of maneuver and intent.

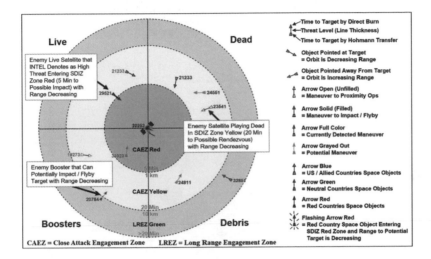

Figure 56: Threat Assessment Charts (TAC) Help Correlate the Meaning of Tactical Activities. *Source:* Adapted from Paul Szymanski, "Issues with the Integration of Space and Terrestrial Military Operations," *Wild Blue Yonder Online Journal* (June 2020). https://www.airuniversity.af.edu/Wild-Blue-Yonder/Article-Display/Article/2226268/ issues-with-the-integration-of-space-and-terrestrial-military-operations/.

The second visualization method employs a Monte Carlo simulation to calculate possible satellite maneuvers based on known location and other orbital parameters. Like the green ellipse in Figure 50, Figure 57 illustrates calculations for threat envelopes surrounding a potentially targeted satellite (red asterisk in center), assuming the attacking satellite is using a low-thrust maneuver profile. Spreading out from the targeted satellite into regions of space are locations (the small circles) from where potential attackers can maneuver to approach their target. The size of each circle corresponds to the amount of Δv fuel required to rendezvous with the target. The color of each circle corresponds to the amount of time it would take to achieve this rendezvous. Thus, higher inclination attacking satellites would need to expend more fuel and take more time to reach the targeted satellite (larger, blue circles). This implies a bigger attacking satellite that can carry more fuel, thus being more easily detected when initiating attack orbital maneuvers, or the defending satellite has more time to detect and prepare for attack. If one can assume maximum sizes of threat satellites, and thus maximum Δv constraints, then these can also be plotted on this chart to give an accurate picture of how much attack warning time is available and what regions of space might require additional surveillance assets to detect such a threat.

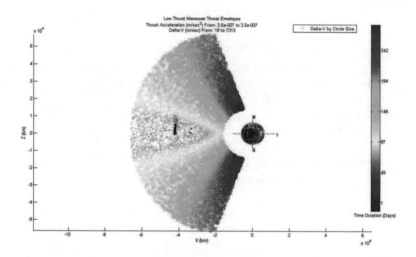

Figure 57: Threat Envelope Maneuver Simulation Centered on a Target Satellite (Red Asterisk). *Source:* Graphic created by Paul Szymanski using the Space Warfare Analysis Tool (SWAT) that he developed.

The black boxes toward the center of the chart are actual locations of current live satellites in relation to the targeted satellite (in flattened space), and the gray boxes denote actual locations of dead space objects. If one assumes that an attack would come from a known, live satellite, then the defending satellite would need to look toward the black boxes for suspicious activities while they are initiating attack. If one assumes the attack would come from a supposedly "dead" space object, then defensive sensors should concentrate on the gray box locations. Otherwise, if one is concerned about an attack coming from "out of the blue," then attack detection sensors must concentrate on successive concentric lower probability fans around the defended satellite based on the assumed size of the attacking object, timeliness of detection, and the defending satellite's timeline for response.

Considering the possibility of attacks in space requires a unique understanding of orbital operations, the potential threats, techniques for their employment, and methods for visualizing and analyzing them. With analysis, the positions of China's BeiDou spacecraft seemed potentially more threatening than would first appear, the unusual behavior of BeiDou 1D raising unique questions about the possibilities of hiding or disguising satellites. From there, it was a small, creative leap to imagine how else one might employ antisatellite weapons on orbit or avoid the sensors trying to search for such systems. These discussions may seem like science fiction, but they are very practical matters for the future of warfare. And there are many more questions to consider across the strategic, operational, and tactical levels of war.

Although this particular vignette takes data that is from over a decade ago, the instance raises significant questions for practitioners of warfare— the fundamental ends of which must be to avoid strategic surprise and to posture our nation and our forces for relative advantage. The ways and means of achieving these ends are myriad and take up much of this book, and although we suggest many, the institutional and the warfighting challenges faced today are significant and complex. Potential adversaries of the United States have differing cultures and approaches to warfare, a fact that can make things seem more complex and obscure when we are dealing with a new theater of operations. Regardless of the domains of conflict, wars will continue to be fought between adversary commanders' minds, and the forces and equipment they employ are simply messaging the opposing commander and influencing his decision making (including willingness to surrender) based on his culture, education, experience, attitudes, fears, emotional and physical energy, and requirements from political leaders. These factors surround the levels of war. As hinted at in the preceding scenario, the coordination of tactical action—and the means by which to determine if tactical action is being coordinated against you—forms the core idea behind the operational level of war and operational art. Along with a preoccupation with visualizing such activities, these are the subjects of the next chapter.

CHAPTER 4

The Language of the Operational Level

Section 1: An Overview of the Schema

The operational level provides the connective tissue that links strategic aims and tactical actions. From the point of view of the strategist, the operational level implements strategy, turning it into activity. From the point of view of the tactician, the operational artist combines tactics with a greater variety of means than is available to the tactician himself. For the operational artist, however, the task is to balance the tension between long-term goals and short-term actions, between the conceptual desires of strategy and the practical constraints of tactics.* For space operations, a schema of the operational level resembles the operational level in other domains, but the seemingly simple elements of the schema belie the complexities of space operations, particularly in visualizing the operational environment, assessing enemy activity, and developing complete plans. While space operators must strive to understand these complexities intimately, all practitioners of joint warfare must be familiar with them to maximize the chances of operational success.

---

* Huba Wass de Czega, "Thinking and Acting Like an Early Explorer: Operational Art Is Not a Level of War," *Small Wars Journal* (2011): 2, https://smallwarsjournal.com/blog/journal/docs-temp/710-deczege.pdf.165.

As with the schemata depicting the strategic and tactical levels, the schema for the operational level is not a complete representation of all activities that occur—only a general shape of them (see Figure 58). It is ultimately the role of the practitioners of space warfare to accept or reject the schema, to develop the procedures and the doctrine to implement the various pieces, and to integrate the theoretical outline with the practicalities of warfighting. That being said, there are certain characteristics of the operational-level schema that look familiar to what appears in the other levels of war.

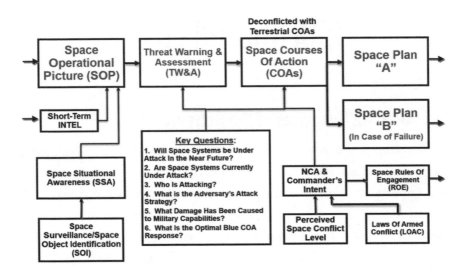

Figure 58: Schema for the Operational Level of War. *Source:* Created by authors.

First, an understanding of the operating environment is essential, including an appreciation for the strategic environment; known friendly, enemy, neutral, and commercial activity; and knowledge of the physical environment. For space operations, these requirements span the orbital, electromagnetic, and terrestrial regimes, making the space operational picture (SOP) a subset of the joint common operating picture. As the arrow leading into the SOP suggests, the operational-level SOP does not exist independently but is a continuously updated product that monitors ongoing operations and accounts for strategic-level activity. The SOP is a current operations tool that employs the military symbology of the previous chapter to visually depict the latest space domain awareness data and finished intelligence. It informs planning, but as the section on visualizing space operations explores, this activity comes with its own unique challenges.

Given an established operating picture, the next required activity is threat warning and assessment. This activity involves not only assessing any ongoing enemy actions but, more important, anticipating what the enemy might do next (an activity called Predictive Battlespace Awareness or PBA). These assessments exist in light of the operating force's understanding of the law of armed conflict, the rules of engagement, and their commander's intent. Several key questions help guide the assessment.

1. Will space systems be under attack in the near future?
2. What space systems are currently under attack?
3. Who is currently attacking, and how?
4. What is the enemy's plan of attack, and how does it fit into their larger strategy?
5. What damage has been caused, and what are the operational impacts?
6. What is the friendly force response?

As these questions suggest, threat warning and assessment is not simply a function of intelligence personnel. Due to its importance, it must involve the whole team, and the results of the assessment will affect the work of the strategist as much as the work of the lawyer and the chaplain. Importantly, the assessment may indicate that the enemy is attempting to escalate the conflict. The following section on the space escalation ladder provides a framework by which to consider the nuanced gradations of space warfare escalation—and therefore the potential for greater control during escalation—than may be possible in other domains. Space provides additional "rungs" on the conflict escalation ladder where countries can express intent and resolve to each other without the general population becoming aware of these actions and forcing their political leaders into terrestrial war. In addition, assuming a potential adversary would attack a country's eyes and ears in space before initiating conflict on the ground, warning of an impending satellite attack may allow military or diplomatic intervention to prevent war from commencing on the earth.

The third major activity within the space operational level schema is the development of space courses of action. Within other frameworks—the joint planning process, for example—much activity centers around the development and analysis of courses of action. The schema presented here does not suggest that those steps are unnecessary, only that within the operational level, the course of action is the operative idea linking the ongoing threat warning and assessment to the development of an executable plan. Importantly, when operating as part of the joint force, space courses of action must complement—not conflict with—the activity occurring within the other domains. This is what is meant by integrating plans in order to synchronize operations. When trying to deny an adversary space capability, one must also coordinate with intelligence agencies to determine whether they might be monitoring electronic traffic through those space systems. The preservation of those intelligence sources might actually be a higher priority than eliminating the space capability altogether.

In addition to the joint planning process (or the military decision-making process for the army or the Marine Corps planning process), other frameworks prove useful in planning space operations. Among these, the elements of operational art and design require particular explanation in light of space operations. The following section on space operational art and design explores these concepts in relation to space warfare in an attempt to develop new ideas based upon existing, established language. To further that goal, the shorter discussions on combined arms dilemmas and stratagems serve as soup-starters for a professional discourse on the creative employment of space warfighting formations.

The final major activity within the operational-level schema is the development of the plan or plans. In simplest terms, we may have a "plan A," which we will execute if all goes "as planned" or a "plan B" to execute if an anticipated situation develops that is not addressed by "plan A." A plan that addresses a secondary contingency is called a branch plan. A plan that follows the execution of another plan is called a sequel. Branches and sequels may be as numerous as time and resources allow but should address most likely and most dangerous scenarios. These plans, of course, synchronize the actions of tactical formations and in so doing link to the tactical level.

It is worth noting also that the development of plans—even if they are never executed—is a useful training opportunity, and the effort to develop plans will certainly pay huge dividends for Space Command as it continues to define its role among combatant commands and its role in the execution of other combatant commands' plans. This problem is more difficult than it may first appear—being prepared to support multiple contingencies across the globe and in space with limited forces. Indeed, for Space Command, every fight is a global fight, and the subsection on mitigating risk through operational planning explores how the balance of resources and risk might be accomplished (more on this in the final chapter).

In building operational-level plans, the operational artist bears responsibility for achieving or partly achieving one or more goals of the strategy with a subset of the strategy's means at his disposal. The major activities in this effort include building and maintaining the operating picture, assessing threat activity, and developing viable courses of action. These activities, in turn—and through the consideration of other frameworks like the space escalation ladder, the elements of operational art and design, and dilemmas and stratagems—lead to a plan, ensuring tactical formations have means at their disposal, are deployed in the right locations on the right timelines, and understand how they will contribute to the operational and strategic goals. With the operational-level schema in mind, the rest of this chapter explores these additional frameworks and their application to operational-level warfare.

Section 2: Visualizing the Synchronization of Tactical Actions

As the operational-level schema suggests, the creation of a common operating picture for space operations (as a subset of the entire joint operating picture) is a crucial activity of space warfare and of war, in general. The process of visualization consists of two complementary but very distinct activities. The first exists in the mind of the commander and is an outgrowth of his understanding and experience, of his ability to make sense of the situation around him and determine how best to proceed. The second activity consists of developing the plan and battle-tracking its execution throughout the operation. This activity is the outgrowth of staff planning processes and the application of technical solutions for visualization. The first activity is the command aspect, the second the control aspect, and neither exists exclusively from the other. In fact, there is give-and-take between the two, the commander's concept of operation and the staff's analysis refining each other. In both activities, however, the common language of symbology and a set of semistandardized views provide a common frame for visualization, one that not only is useful for space operations commanders but is also translatable to other joint force commanders in the planning and execution of all-domain operations so that even nonspace personnel can see how space operations impact the rest of the battlefield.

Chapter 2 painstakingly laid out examples of military symbology, both extant and emergent, as a foundation for such communication. From an institutional perspective, these symbols—or thematically similar ones—require adoption by the training programs for service members of all disciplines. In this way, the landsman, the seaman, and the airman may gain a base level of functionality with space domain operations, and the space operator may acquire the building blocks necessary for employing tactical activity—as illustrated in some of the following graphics. In a similar way, a few standard views of space operations—both in planning and in execution—provide mental models for what products may look like throughout an operation. This approach, it should be noted, is not different than planning for operations in other domains. The detailed plan proceeds from the concept, and different ways of visualizing support different activities within the planning process. Conceptual graphics may take on the form of scheme-of-maneuver or synchronization sketches. Oppositely, real-time battle-tracking graphics may appear in a variety of two-dimensional plots or three-dimensional depictions, which are sometimes overlaid on maps or globes.

The Conceptual Visualization

Conceptual visualizations are akin to sketches or cartoons that deliberately oversimplify a scheme of maneuver or a course of action. This level of planning provides a springboard for detailed planning, without which a thorough plan cannot come to fruition. Importantly, planners may go back and forth between conceptual and detailed planning, the conceptual suggesting areas for more detailed investigation, the detailed investigation revealing difficulties or shortcomings with the original concept or specific portions of it. A detailed analysis of a satellite's onboard fuel, for instance, may reveal that the original concept for orbital maneuver is not acceptable because it will dramatically shorten the life expectancy of the satellite. As a result, it will be necessary either to revise the maneuver concept or to accept that the mission is important enough to sacrifice the satellite.

The following conceptual visualization provides an example of an enemy's general scheme of maneuver—a product that an intelligence shop may brief for a high-level visualization of an operation (Figure 59). It is a concept only—not a complete plan—and it must have additional detailed planning behind it to be viable. The following cartoon includes a map of the earth as its base. In this notional scenario, friendly forces have been conducting orbital space domain awareness operations to determine the nature of an adversary's newly launched satellite. The recent introduction of jammers into the theater, the presence of a known kinetic kill vehicle satellite, and other intelligence indicators suggest that the enemy's scheme of maneuver will involve coordinated cross-domain and in-domain attacks against friendly assets to allow the new enemy satellite freedom of maneuver. Based on this intelligence picture, friendly forces can begin developing a concept for how they might respond.

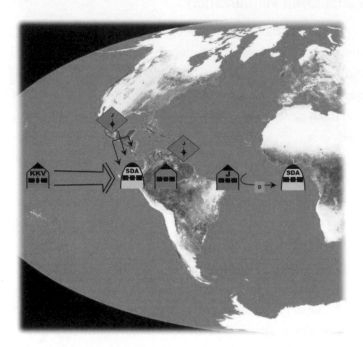

Figure 59: Conceptual Visualization of an Enemy's Scheme of Maneuver. *Source:* Created by authors with adapted base map from "Global Vegetation," NASA, last modified April 7, 2008, https://www.nasa.gov/multimedia/imagegallery/image_feature_1056.html.

Synchronization Matrix Sketch

The synchronization matrix sketch also illustrates a conceptual depiction of space operations activity at the operational level but enjoys several distinct characteristics from the preceding example (see Figure 60). First, the synchronization matrix visualization includes a consideration of other domain forces (in this case, primarily land forces) and the operational timing associated with those domains. Second, the following sketch is less concerned with the orbital warfare aspect of space operations than with the traditional space support missions like missile warning (MW); positioning, navigation, and timing (PNT); satellite communications (SATCOM); intelligence collection (IC); and defensive space control (DSC). The ups and downs of the colored lines in rows 9 through 11 indicate the availability of space support assets. In Row 9, for example, air force missile warning assets remain on mission throughout the operation, whereas assets in Central Command come off of mission on G-Day (the day the ground operation begins), and assets in European Command come onto mission that day. Rows 10 and 11 chart the anticipated strength of PNT and SATCOM signals throughout the operation, and Row 12 depicts windows of observation by friendly, enemy, and commercial collection assets. Row 13 functions like the missile warning row, indicating that defensive space control assets in the Pacific will not be available after G+2. Finally, Row 14 lists some options to make space support more effective.

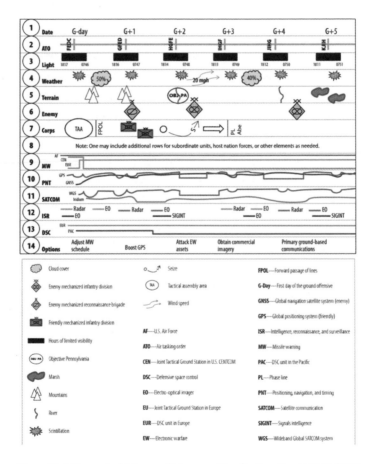

Figure 60: A Conceptual Scheme of Maneuver Visualization. *Source:* Jerry Drew, "Visualizing the Synchronization of Space Systems in Operational Planning," *Military Review* (January–February 2019): 106–114. https://www.armyupress.army.mil/ Journals/MilitaryReview/English-Edition-Archives/Jan-Feb-2019/Drew-Space/.

In the operations of the joint force, conceptual visualizations and synchronization visualizations can both serve their purpose, the level of detail on the sketches increasing as the plans develop. Likewise, while visualizations of orbital warfare may certainly exist independently, one may easily imagine additional rows on the synchronization matrix sketch to account for the activities of orbital assets or ground-based jammers. With the necessary conceptual visualizations in hand, one may begin considering more detailed visualizations.

Detailed Visualizations

Whether for planning, preparation, execution, or assessment, detailed visualizations may come in two-dimensional or three-dimensional forms, using standardized symbology. If the preceding conceptual visualizations more accurately reflect the command function of visualization, the more detailed visualizations that follow are better suited for the control function. The visualizations here leverage a computer program developed by Mr. Szymanski called Space Attack Warning (SAW) and its organic symbology.* Figure 61 shows a view that would be familiar to space operations personnel with the various orbits depicted as ground traces. What makes this view different, however, is that it employs standardized symbology for both satellites and ground-segment space operations units, and it also color codes the ground traces as blue (friendly), red (enemy), green (neutral), or unknown (yellow) in accordance with the standard rules of military symbols. This view is particularly beneficial in the development of space surveillance plans, understanding attack envelopes, and supporting wargame events.

---

* For more information on SWAT/SAW products or for analysis, contact Paul Szymanski via LinkedIn at https://www.linkedin.com/in/paulszymansk.

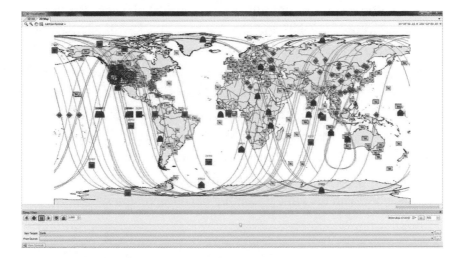

Figure 61: SAW Flat Map View. *Source:* Graphic created by Paul Szymanski using the Satellite Attack Warning (SAW) software developed by the Metatech Corporation under his direction.

Through the filtering of data, one may take the same information and view it as a plot of satellite altitudes versus satellite inclinations—the so-called Altitude versus Inclination Survey, or AVIS plot. These two factors are the most important in determining the feasibility of operational maneuvers—how far and how fast the satellite can move by changing its velocity with the amount of on-board fuel.* In plotting this information, Figure 62 illustrates the concept of space choke points, showing the tendency of space objects to concentrate in certain orbits. This particular chart shows all space objects, noting three typical orbits—low-earth orbit (LEO), medium-earth orbit (MEO), and geosynchronous orbit (GEO). It simplifies the space situation view for the warfighter by only illustrating orbital or inclination changes, fixing the actual orbital movement and instead depicting the satellites in their positions relative to one another.

In the following chart (Figure 62), each dot represents an individual space object. In the altitude-inclination space, close dots do not necessarily indicate closeness in orbital space. One satellite (an attack satellite) may be on the opposite side of the earth from another satellite (its target), but because

---

*   In orbital mechanics, the change in velocity is referred to as Δv or "Delta Vee."

its altitude and inclination are similar, it is considered "close" because the attacker can choose the time and phasing of its attack with relatively little fuel expenditure. This chart assumes that the method of attack is a rendez-vous-and-proximity operation (RPO)—a type of attack where the attacker essentially matches the orbit of the target (a co-orbital attack). For the case of a glancing attack where the orbits do not match but the orbital tracks intercept (for example, the collision between COSMOS 2251 and Iridium 33, which had a 12-degree difference in inclination but the same orbital altitude), this AVIS chart should show all objects of the same altitude as a horizontal line with the same possibility of collision/attack.*

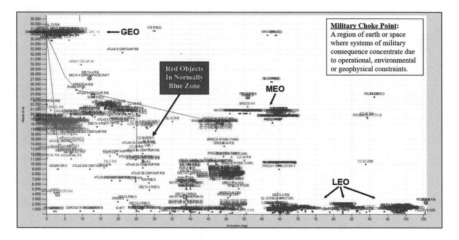

Figure 62: SAW—All Altitudes. *Source:* Graphic created by Paul Szymanski using the Satellite Attack Warning (SAW) software developed by the Metatech Corporation under his direction.

One can see in Figure 62 that certain altitude-inclination regions have a preponderance of one country's space objects over another's. These are examples of choke points, and they are significant for operational-level analysis. For example, there is a density of red-force objects at 50 degrees/8,000 kilometers (low center on the graph, probably due to the latitude of this country's launch sites). When a potential threat country's object appears in

---

*   Becky Iannotta and Tariq Malik, "US Satellite Destroyed in Space Collision," Space.com, February 11, 2009, http://www.space.com/news/090211-satellite-collision.html.

a zone that is predominantly blue, or when a satellite has maneuvered over a user-defined time span, further analysis to explore the potential threat is likely required. The lines in this graph show how two satellites (one friendly and one neutral) have moved from geosynchronous transfer parking orbits to geosynchronous ones (from bottom right to top left). Locations where red and blue satellites routinely intermingle—as in the region around 97 degrees/600 kilometers, a valuable spot for sun-synchronous imagery satellites—provide another area ripe for threat activity analysis.

Figure 63 zooms into the region near 97 degrees/600 kilometers. In addition to the notation of each object's country and mission, data points that are much more easily viewed at this level of zoom, trailing lines denote the orbital history of the objects over an operator-designated time period. The red object "JB-2C" has executed an orbital altitude increase over the time period and is easily identified in this view moving away from neutral orbital debris, a prudent measure. If, however, the trailing lines appeared to denote that the satellite was moving toward another satellite, this may imply that the satellite is posturing for an attack.

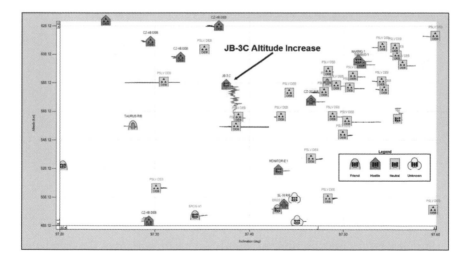

Figure 63: SAW View, LEO. *Source:* Paul Szymanski, "Issues with the Integration of Space and Terrestrial Military Operations," *Wild Blue Yonder Online Journal* (June 2020). https://www.airuniversity.af.edu/Wild-Blue-Yonder/Article-Display/Article/2226268/ issues-with-the-integration-of-space-and-terrestrial-military-operations/.

During periods of relative peace, avoiding orbital debris is imperative for the continuation of space operations. During combat operations, when the potential for the creation of new debris is increased, predictive analysis of debris fields will be as useful in planning as avoiding such debris is during current operations. Figure 64 provides an illustration of the debris cloud formed following the breakup of the Russian satellite COSMOS 2421 at an altitude of approximately 392 kilometers (the red icon). The yellow icons indicate major debris fields generated by the breakup, but due to limitations of SDA sensors, it would not be possible to track every small piece at all times. Thus, predictive analysis is necessary. The squiggly lines denote the gradual decay from orbit of these debris objects based on somewhat uncertain space surveillance orbital parameters.

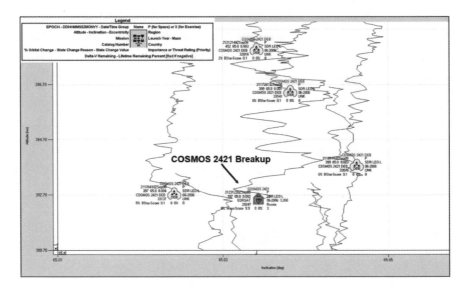

Figure 64: Visualization of a Debris Cloud. *Source:* Graphic created by Paul Szymanski using the Satellite Attack Warning (SAW) software developed by the Metatech Corporation under his direction.

Supplementary information surrounds each icon and gives the data date-time group, country of origin, mission, launch date, mass, specific orbital parameters, the space defense region this object occupies, its space catalog number, and a SWAT-calculated degree of change score. In this scale, a 9.9 indicates that this object has experienced a significant change in one or more of its twenty-six measurable space domain awareness parameters (orbit, size, shape, radar cross section, optical cross section, and so on) along with the main reason for the state change (in the case of COSMOS 2421, its mass).* As one might expect, the breakup of a satellite would lead to significant change of its tracked properties. Also, as one might expect, the score for the newly created orbital debris fields is very low since their parameters are only being defined after the breakup of the parent space-craft. These debris clouds are expected to follow the zigzagged lines, many of them drifting into higher altitudes. In this view, the movement of the debris masses does not approach any other object—a good thing because more collisions would not only endanger other satellites but create more orbital debris.

As discussed previously, the idea of dividing the entirety of space into multiple operating areas has great utility. In the case of an orbital debris threat, the organization responsible for each operating area can focus on the protection of its assets. In the case of intentional attack, such delineation of responsibilities ensures a focused and coordinated response.

The SAW view in Figure 65 shows space defense identification zones (SDIZ). Akin to air defense identification zones, the SDIZ provides an organizing principle for space operations and help in managing space control assets and in interpreting indications and warnings. In the particular conception shown in Figure 65, the largest SDIZ (bottom left) is named "Blue Zone—LEO," which is in adjacent to "Red Zone—MEO," "Red Zone—MEO-Low," and "Red Zone—LEO-High 1." The satellites in these red zones may not be physically close to any blue zone satellites at a particular moment. However, because of their potential to maneuver from complementary altitude/inclination positions, they could feasibly pose a threat. As the commander responsible for the entirety of the space area of operations, the commander of Space Command would necessarily be responsible for understanding the reports coming from the subordinate commanders responsible for these SDIZs and in developing a coordinated response in intelligence collection, space surveillance sensor

---

*   In this method of calculation, values above seven generally mean major changes to the satellite's historical characteristics.

tasking, changes in weapons release authorizations for satellite self-defense, and other aspects of the operation.

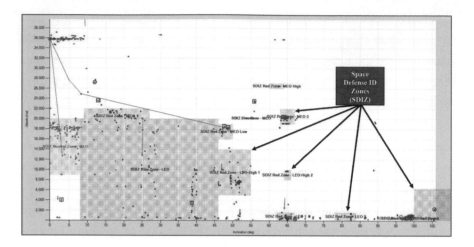

Figure 65: Space Defense Identification Zones. *Source:* Graphic created by Paul Szymanski using the Satellite Attack Warning (SAW) software developed by the Metatech Corporation under his direction.

Figure 66 illustrates how a simulated attack against GPS satellites might look. In the center of the display, multiple expended "dead" boosters of enemy (red) launch vehicles appear to be maneuvering toward a GPS satellite. The historical activity of these boosters has always shown them to be outside of altitudes and inclinations that were threatening to GPS. However, with this new activity, the boosters are entering a designated blue SDIZ. The question remains: is the enemy engaging in deliberate maneuver, or is this the result of a drifting or malfunctioning piece of space junk? Other activity in the space domain (or indeed in other domains) may provide clues and additional indicators of massed space attacks both on orbit and on the ground.

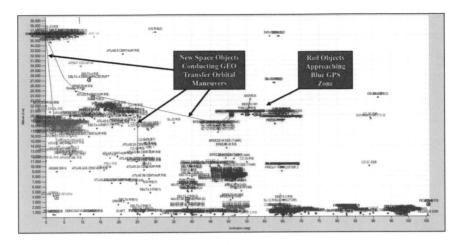

Figure 66: Simulated Attack against GPS. *Source:* Graphic created by
Paul Szymanski using the Satellite Attack Warning (SAW) software
developed by the Metatech Corporation under his direction.

In addition to at least one red object approaching the GPS SDIZ, three
new objects are conducting maneuvers that will allow them to transfer
from their current orbits into a GEO orbit. The first transfer is a commercial satellite exiting a heavily commercial zone (green, far left). In itself,
this activity does not seem suspicious, but it should not be discounted as
part of a coordinated effort, especially if the commercial entity operating
the satellite has ties with an adversarial nation. The second transfer (blue)
is known to friendly forces and has been anticipated for quite some time.
Depending on the nature of this satellite, the adversary's activity may be
partially in response to this maneuver. However, the adversary may simply
be conducting a maneuver of his own accord, and we may be trying to
construct a narrative that does not rightly exist. The final maneuver is a
red satellite near 25 degrees of inclination. Its inclination is not near the
suspicious booster, but this satellite may be part of a coordinated effort to
observe the booster's maneuver or to threaten a secondary target, or it may
be in pursuit of an entirely different mission. Certainly, making sense of
what is happening in the orbital battle is no easy task.

Upon closer study, more details reveal themselves. In this simulation,
multiple "dead" rocket boosters have unexpectedly come back to life and are
maneuvering in a threatening manner. Figure 67 zooms in on the previous

chart and shows the orbital changes of the red boosters by the trace lines leading toward GPS satellites. As the seven antisatellite weapons converge in what appears to be a coordinated attack, friendly forces must begin generating options for the survivability of GPS as a constellation. With the loss of potentially seven GPS satellites—assuming each ASAT can destroy only one friendly satellite—the capability of the constellation and its ability to support land, air, maritime, civil, and commercial activity could be severely jeopardized. Indeed, degradation would have serious impacts not only to the joint fight but also to the global economy.

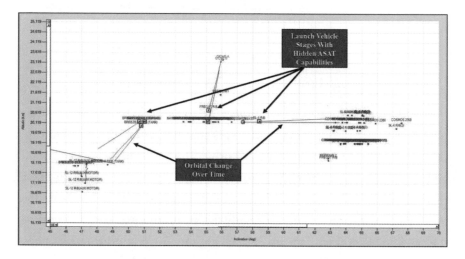

Figure 67: Coordinated Attack against GPS Satellites (Notional). *Source:* Paul Szymanski, "Issues with the Integration of Space and Terrestrial Military Operations," *Wild Blue Yonder Online Journal* (June 2020). https://www.airuniversity.af.edu/Wild-Blue-Yonder/Article-Display/Article/2226268/issues-with-the-integration-of-space-and-terrestrial-military-operations/.

Figure 68 zooms in even further, showing the final approach of the attacking spacecraft. By now, the joint force has caught on to the attack and reoriented its space domain awareness sensors to observe what is happening. In the amplifying information of each enemy icon, the state change scores are large because the radar cross sections (RCS) of each ASAT have radically changed, reflecting the changes in attitude of the rocket bodies as they orient themselves in coordinated attack.

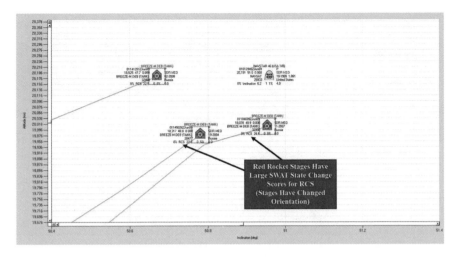

Figure 68: The Final Assault. RCS Numbers Are High, Indicating Recently Changed Orientations. *Source:* Graphic created by Paul Szymanski using the Satellite Attack Warning (SAW) software developed by the Metatech Corporation under his direction.

To be sure, a large amount or uncertainty exists about the nature of numerous orbital objects. It is possible, however, based on known bits of information, to piece together reasonable conclusions about the nature of these objects. Again, we rely on the analytical capability built into the SWAT software. The upper left table shown in Figure 69 lists the possible missions of Resident Space Object-0035 (RSO-0035). The boxes across the bottom of the screen house information from three broad categories: Satellite Description, Optical Properties, and Two-Line Element Set. Each of these broad categories is further broken down into a series of indicators. For example, in the "Satellite Description" box, we have information about RSO-0035's drift rate, RCS value, and mass, each of which is given a score. Based on the number of indicators available across the three broad categories and the weights of those indicators, the most likely mission (77 percent likely) for RSO0035 is assessed to be commercial communications.

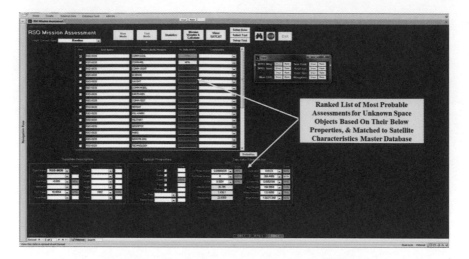

Figure 69: Automatic Assessments of Unknown Space Objects' Most
Probable Missions. *Source:* Graphic created by Paul Szymanski using the
Space Warfare Analysis Tool (SWAT) developed by himself.

In space warfare, it is equally necessary to know about the state of the
enemy's spacecraft as it is to know how those states are changing. Based on
a regularly updated list of satellite parameters, the automated data sorting
capacity of SWAT allows planners to determine which satellites have under-
gone the most significant state changes and what such changes may mean
for the state of the conflict (Figure 70). Recall the earlier vignette about
China and the oddities of its BeiDou constellation. An automated analysis
like this would show when a number of BeiDou satellites were reorienting,
relocating, or emitting different signatures than those that are considered
nominal. Were that to happen, we may infer that they were preparing to
mass on the GPS constellation within some particular choke point.

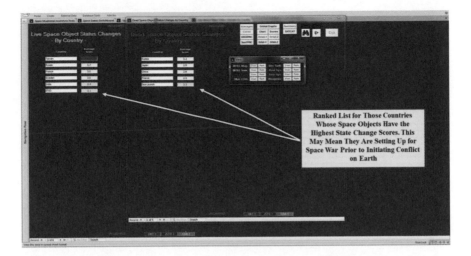

Figure 70: Automatic Assessment of Multiple Space Object State Changes. *Source:* Graphic created by Paul Szymanski using the Space Warfare Analysis Tool (SWAT) developed by himself.

With the most probable mission established and some knowledge of their spacecraft state changes, it is next possible to assess other factors of about the enemy's course of action. First, SWAT's database uses a list of intelligence indicators to inform this process. In the space warfare scenario shown in Figure 71, a number of indicators are being considered to assess the likely tactical employment of the enemy's systems, including the employment of a mobile laser blinder (most likely), mobile direct-ascent ASATs, maintenance satellites, mobile ground jammers, and nanosatellite paint mines in LEO. These tactics combine with other indicators to suggest that the enemy may be preparing to employ a "Sweep the Skies" stratagem (see Chapter 4, Section 6 for more on stratagems) as the conflict moves into a transconflict stage.

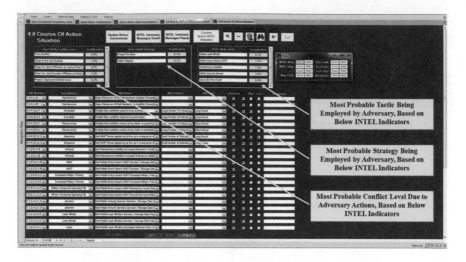

Figure 71: Automatic Operational Assessments of Active Space Objects'
Most Probable Intents. *Source:* Graphic created by Paul Szymanski using the
Space Warfare Analysis Tool (SWAT) that he developed by himself.

Three-Dimensional Views

For orbital operations, two-dimensional views such as those depicted here lend themselves more to analysis than their three-dimensional counterparts. The primary benefit of the three-dimensional views, however, is that they provide a more intuitive depiction of the operating environment, and for limited scope operations, these views may be more helpful. Figure 72 shows a 3-D depiction of orbital activity over North America. The inclusion of icons over the globe, along with select orbital tracks, hints at the possibility of interpretation, but even skilled operators would have difficulty determining the meaning of activities at this scale. It would be possible to assign operators focused portions of this SOP, allowing them to watch for certain activities or indicators with a fusion cell collating observations and piecing together the meaning of the bigger picture.

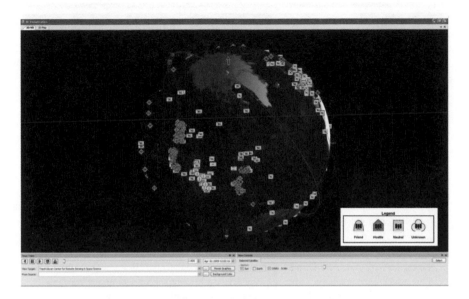

Figure 72: 3-D Visualization of Notional Orbital Activity over North America.
Source: Graphic created by Paul Szymanski using the Satellite Attack Warning
(SAW) software developed by the Metatech Corporation under his direction.

Figure 73 is slightly zoomed in and includes sensor "bubbles" for ground-based sensors. This view suggests the multidomain view of space operations insofar as it includes a consideration of some surface-based capabilities that are emitting electromagnetic radiation. Importantly, with the inclusion of ground sensors, we can see where sensor coverage is lacking. Since activity is not observable within these dead zones, an adversary may take advantage of the opportunity to alter the orbit of their spacecraft or to conduct in-flight testing out of view. In turn, it is incumbent upon the friendly force to reacquire the spacecraft emerging from the dead zone as quickly as possible to confirm or deny the change of its orbital parameters and if they have changed, to determine what that means for the overall operation.

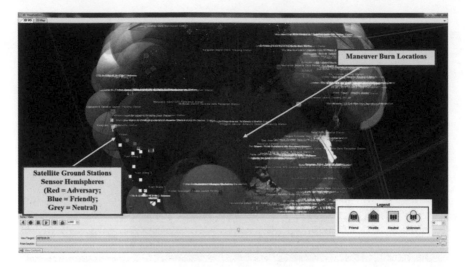

Figure 73: Orbital Activity, Ground Sensor Coverage, and Potential Locations for Maneuver for Attack. *Source:* Graphic created by Paul Szymanski using the Satellite Attack Warning (SAW) software developed by the Metatech Corporation under his direction.

Whatever the technique of visualization employed, the goal is the same: to increase the likelihood of shared understanding across the force in a timed-phased and terrestrial-linked manner. Shared understanding, whether in planning or in execution, will assist in successful achievement of designated objectives, and for the operational level, achieving objectives directly links to the achievement of a strategic aim. Throughout the execution of the operation, however, the interplay of friendly and enemy forces will lead to a series of actions and counteractions, some of which may alter the understanding of the situation dramatically—possibly even altering the strategic aims themselves. To think through how the actions and counteractions of space operations may escalate tensions, we introduce the space conflict escalation ladder.

Section 3: The Space Conflict Escalation Ladder

Perhaps the most important addition to military strategy that space warfare offers is a nuanced menu of alternatives. These alternatives provide the opportunity for the nation's highest decision-makers to employ space power assets to seek political objectives rather than employing more overt or more disruptive options. In short, maneuvering a satellite near an adversary nation's satellite may be equally effective at messaging displeasure with an adversary's actions on earth but is nowhere near as provocative as anchoring an aircraft carrier just outside of their economic exclusion zone. That being said, there are times when only an aircraft carrier will do, but even in those situations, space operations are taking place in the background to aid in the success of the fleet and other forces.

The following space conflict escalation ladder presupposes an agreed-upon phasing construct that numbers phases sequentially. This numbering scheme does not follow the legacy five-phase numbering scheme and will not be valid for all conflicts. It does, however, offer numerous suggestions for employment options available across the conflict continuum, even suggesting subphases (for example, 4.A and 4.B) and activities within those subphases (4.B.1, 4.B.2). In this model, as the phase number grows larger, the conflict escalates, moving from cooler to hotter colors as the conflict matures from preconflict to transconflict to postconflict. Importantly, the phases of the space conflict in this depiction do not always coincide with the phases of the terrestrial campaign. It is possible that the terrestrial conflict escalates while the space conflict plateaus or vice versa. It is also possible that conflict escalation in one sphere might cause escalation or de-escalation in another. The following depiction, then, only suggests one possible way, and it is left to future planners to modify for their purposes.

The escalation ladder begins in the preconflict buildup (Figure 74). Characteristic of this phase (P.1) is the anticipation of imminent conflict and the necessity of preparations. Activities may include overt weapons testing and deployment, the modification of international treaties or support agreements, and ramped-up reconnaissance efforts. As we transition into the preconflict crisis (P.1.C), tensions are rising. If the conflict escalates linearly, we may expect continued mobilizations in multiple domains, increased efforts to employ covert and deceptive techniques and technologies, and more aggressive acts. If the enemy does not act in the way that our model suggests, however, or if he chooses to escalate in order to de-escalate, the linear suppositions may no longer apply.

WBS	Conflict Phase	Terrestrial Campaign Phase	Space Campaign Phase	Weapon Type
P.1.A.0	Preconflict	Phase 0: Prewar Buildup (Shape)	1st Wave Attacks Phase A	Preconflict Deter
P.1.B.0	Preconflict	Phase 0: Prewar Buildup (Shape)	1st Wave Attacks Phase B	Persuade; Spying; Propaganda; Avoidance Maneuvering; Increased Space Surveillance & Close Satellite Inspections
P.1.C.0	Preconflict Crisis	Phase 0: Prewar Buildup (Shape)	1st Wave Attacks Phase C	Hide; Covert; Cyber; Political Disruptions; Mobilize Forces; Increase Military Alert Level; Threatening Satellite Maneuvers; Increase Space Radiation; Initiate Satellite Defensive Measures; Employ Nation's Astronauts on International Space Station for Military Uses
P.2.A.0	Transconflict	Phase I: Deployment/ Deterrence (Deter)	2nd Wave Attacks	Transconflict Deter

Figure 74: Space Escalation Ladder (Early Conflict). *Source:* Adapted from Paul Szymanski, "Issues with the Integration of Space and Terrestrial Military Operations," *Wild Blue Yonder Online Journal* (June 2020). https://www.airuniversity.af.edu/Wild-Blue-Yonder/Article-Display/Article/2226268/issues-with-the-integration-of-space-and-terrestrial-military-operations/.

Space Campaign Phase Full Name	Weapon Category
1st Wave Attacks Phase A - Preconflict Deter	Overt Weapons Testing & Deployment; Treaties; Saber Rattling; Space Alliances; Normal Space Surveillance, Tracking & Reconnaissance Activities; Satellite Close Inspectors
1st Wave Attacks Phase B - Preconflict Persuade	Diplomatic Requests & Démarches; Economic Actions; Embargos; Legal Actions; Administrative Actions; Transmitting Propaganda Broadcasts; Jamming Propaganda Broadcasts; Increased Spying & Surveillance; Unusual Increases in Space Surveillance and Tracking Activities; Threaten Allies of Your Adversaries; Maneuver to Avoid Attacks
1st Wave Attacks Phase C - Preconflict Hide	Camouflage; Stop Activities; Mobility; Covert Technology Developments; Small Covert SOF Attacks; Cyberattacks; Covert Actions in Violation of International Treaties; Cut off Diplomatic Relations; Inspire Social Disruptions and Agitation; Employ Lethal Force against Your Own Citizens; Mobilize Forces; Increase Military Alert Level (DEFCON); Maneuver Close Enough to Adversary Satellites to Purposely Appear as a Threat; Reveal Covert Programs to Appear Threatening; Enter Into War-Reserve Modes (Hide) for Critical Satellites; Hide Senior Leadership; Increase Radiation Environment in Orbits Used by Adversaries; Initiate Satellite Defensive Measures; Employ Nation's Astronauts on International Space Station for Military Reconnaissance and Surveillance; Spoof and Falsify Worldwide Distribution of Satellite Location Orbital Tracking Data
2nd Wave Attacks - Transconflict Deter	Provocative but False Attacks; Linked Attacks; Demo Attacks; Alternate Country Attacks; Blockades; Major Covert SOF Attacks; Terrorist Attacks; Summarily Execute Saboteurs; Seize & Sequester Suspected Terrorists; Alert Antisatellite Systems; Arm Satellite Self-Defense Mechanisms; Alert Antimissile Defenses; Alert Antiaircraft Defenses; Arm Allied Astronauts on International Space Station

Once the conflict has begun, the rules of engagement and an intimate understanding of the potential effects of space activities become exceedingly important. To that end, the primarily tactical considerations of weapons release authorities and allowed changes to activities as the conflict escalates are among the most significant portions of the plan and must be understood by all. The "Troubled Peace" vignette between Chapters 4 and 5 addresses the specifics of how models for weapons release authorities and rules of engagement may look, but for the purposes of the discussion on the escalation ladder, generalities will suffice. In general, and as depicted in the escalation ladder, weapons release authorities become more permissive as the conflict escalates. For example, while disruption and denial may be the only activities authorized early on, degradation and destruction may become more acceptable to decision-makers eager to put an end to the war or to limit the adversary's ability to wage the next war (see Figure 75).

WBS	Conflict Phase	Terrestrial Campaign Phase	Space Campaign Phase
P.3.A.1	Transconflict	Phase II: Halt Incursion (Seize Initiative)	3rd Wave Attacks Phase A1—Ground Based
P.3.A.2	Transconflict	Phase II: Halt Incursion (Seize Initiative)	3rd Wave Attacks Phase A2—Ground Based
P.3.B.1	Transconflict	Phase III: Air Counteroffensive (Dominate)	3rd Wave Attacks Phase B1—Space Based
P.3.B.2	Transconflict	Phase III: Air Counteroffensive (Dominate)	3rd Wave Attacks Phase B2—Space Based
P.4.A.1	Transconflict	Phase IV: Joint Counteroffensive to Restore Friendly Preconflict Status (Stabilize Borders)	4th Wave Attacks Phase A1—Ground Based
P.4.A.2	Transconflict	Phase IV: Joint Counteroffensive to Restore Friendly Preconflict Status (Stabilize Borders)	4th Wave Attacks Phase A2—Ground Based
P.4.B.1	Transconflict	Phase V: Joint Counteroffensive to Capture Adversary Capitol (Enable New)	4th Wave Attacks Phase B1—Space Based
P.4.B.2	Transconflict	Phase V: Joint Counteroffensive to Capture Adversary Capitol (Enable New)	4th Wave Attacks Phase B2—Space Based
P.5.A.0	Transconflict	Phase VI: Defend against Adversary Counterattacks against Friendly Homeland (Defend Friendly Citizens)	5th Wave Attacks
P.6.A.0	Transconflict	Phase VI: Defend against Adversary Counterattacks against Friendly Homeland (Defend Friendly Citizens)	6th Wave Attacks
P.7.A.0	Transconflict	Phase VII: Defend against Adversary Use of Nuclear Weapons in Space (Defend Friendly Military)	7th Wave Attacks
P.8.A.0	Transconflict	Phase VIII: Defend against Adversary Use of NBC Against Friendly Military Targets (Defend Friendly Military)	8th Wave Attacks; Phase A—Military Targets
P.8.B.0	Transconflict	Phase IX: Defend against Adversary Use of NBC against All Friendly Targets (Defend Friendly Military & Civilians)	8th Wave Attacks; Phase B—Civilian Targets
P.9.A.0	Postconflict	Phase X: Posthostilities (Reconstruction & Stabilization)	9th Wave Attacks

Figure 75: Space Escalation Ladder (Late Conflict). *Source:* Adapted from Paul Szymanski, "Issues with the Integration of Space and Terrestrial Military Operations," *Wild Blue Yonder Online Journal* (June 2020). https://www.airuniversity.af.edu/Wild-Blue-Yonder/Article-Display/Article/2226268/issues-with-the-integration-of-space-and-terrestrial-military-operations/.

Weapon Type	Space Campaign Phase Full Name	Weapon Category
From Terrestrial Partial Temporary Kill	3rd Wave Attacks Phase A1— Terrestrial-to-Space Partial Temporary Effects	Delay, Deny, Covertly Assassinate Adversary Diplomatic Ambassador
From Terrestrial Total Temporary Kill	3rd Wave Attacks Phase A2— Terrestrial-to-Space Total Temporary Effects	Disrupt
From Space Partial Temporary Kill	3rd Wave Attacks Phase B1— Space-to-Space Partial Temporary Effects	Delay, Deny
From Space Total Temporary Kill	3rd Wave Attacks Phase B2— Space-to-Space Total Temporary Effects	Disrupt
From Terrestrial Partial Permanent Kill	4th Wave Attacks Phase A1— Terrestrial-to-Space Partial Permanent Kill	Degrade
From Terrestrial Total Permanent Kill	4th Wave Attacks Phase A2— Terrestrial-to-Space Total Permanent Kill	Destroy
From Space Partial Permanent Kill	4th Wave Attacks Phase B1— Space-to-Space Partial Permanent Kill	Degrade
From Space Total Permanent Kill	4th Wave Attacks Phase B2— Space-to-Space Total Permanent Kill	Destroy
Space-Manned Permanent Kill: Kill Adversary Astronauts	5th Wave Attacks— Space-Manned Permanent Kill	Degrade, Destroy: Kill Adversary Astronauts on International Space Station
Space-to-Earth Permanent Kill	6th Wave Attacks— Space-to-Earth Permanent Kill	Degrade, Destroy
NBC Use—Space	7th Wave Attacks— NBC Use—Space	Degrade, Destroy
NBC Use—Space & Terrestrial	8th Wave Attacks Phase A— NBC Use—Space & Terrestrial— Military Targets	Degrade, Destroy
NBC Use—Space & Terrestrial	8th Wave Attacks Phase B— NBC Use —Space & Terrestrial— Civilian Targets	Degrade, Destroy
Postconflict Deter	9th Wave Attacks— Postconflict Deter	Diplomatic Requests; Economic Actions; Legal Actions; Administrative Actions; Jamming Propaganda Broadcasts

Among all possible choices in war, nuclear options remain on the table. Phases 7 and 8 of the preceding model consider this potentiality, which takes on an interesting twist for space warfare. A belligerent who is not as dependent on space (North Korea, for example) may be willing to launch a nuclear weapon into space, either detonating it or placing it into orbit for later deterrence or employment. The intent of such an action would be to render a key portion of their enemy's satellite fleet inoperable or large portions of space unusable, thus inducing friction (or even operational shock) into the campaign and opening space to negotiate the end of the conflict. This tactic may provide an example of escalating to de-escalate; however, the use of a nuclear weapon, even in space, may provoke any number of responses. One might even consider the possibility that if a country is willing to suffer the consequences of detonating nuclear weapons, space might not be the best place to do so because of the potential for diminished political or psychological impact. After all, there will be no smoking hole in the ground afterward.

Whatever the chosen response option, it is best considered beforehand, and the space escalation ladder offers a framework for considering how actions may play out. For that reason, it is a useful tool for developing and war-gaming plans, for deconflicting activities among services, and potentially for battle-tracking activities during the space wars of the future (that is, as an execution matrix). Perhaps more importantly, the operational artist could use a framework like this to convey to a joint force commander—possibly one unfamiliar with space operations—the litany of options that are available across the conflict of continuum, what means may enable those options, and what outcomes may possibly result. Of course, perceptions of conflict escalation limits are very subjective and culturally dependent, and misinterpretations of low-threat military action could conceivably lead to increased levels of conflict. It may be difficult to distinguish between a deliberate attack and a routine system failure as in the case of the Russian failures during the Crimean conflict. It may be difficult for a belligerent to understand why a particular satellite or ground station is under attack, particularly if it is not currently engaged in the theater. Add the obscurity of on-orbit and electromagnetic spectrum operations and a limited number of historical experiences from which to draw in these areas, and uncertainty mounts.

Section 4: Elements of Space Operational Art and Design (SOAD)

Operational art and design are fundamental ideas to the operational-level schema of space operations. Definitionally, operational design is the process by which joint commanders and staffs construct frameworks for campaigns, and operational art is the cognitive approach used to "develop strategies, campaigns, and operations."* The distinction between the two ideas is that operational design is a subset of operational art. Design allows us to make sense of the current situation and to develop a broad approach that will lead us to our desired future state. We are conducting operational art during design just as we are conducting it after design when we begin more detailed planning. To achieve these functions, both operational art and operational design rely on a list of elements—points of consideration that, when applied, allow a military planner to focus on the most relevant aspects of an operation. Table 32 shows the joint elements of operational design alongside the army's elements of operational art. Note that, while there is some overlap, the elements are not the same.

Often tied to campaign planning, these elements are most commonly adopted at the operational and strategic levels, but like the principles of war, the joint elements of operational design and the army's elements of operational art provide concepts that are useful across the levels of war.† Since the elements of operational design are slightly different than the elements of operational art, it is worth addressing the concepts through doctrinal definitions and historical examples, asking why there are such differences and exploring, through hypothetical scenarios, what new considerations the elements require in light of space warfare. Distinct as they are, both sets are useful, and because space operations inherently span the joint force, we consider the sets together as elements of space operational art and design.‡

---

*   US Joint Staff, Joint Publication (JP) 5-0, *Joint Planning* (Washington DC: Government Printing Office, 2017), IV-1.

†   As Joint Publication 5-0 states, the "elements of operational design can be used for all military planning," which implies the utility of the concepts across the levels of war (IV-19).

‡   We follow the order of the elements of operational design as listed in Joint Publication 5-0 and the order of elements of operational art as listed in Army Doctrine Reference Publication 5-0.

Elements of Operational Design (Joint Force)	Elements of Operational Art (Army)
Termination	
Military End State	End State and Conditions
Objectives	
Effects	
Center of Gravity	Center of Gravity
Decisive Points	Decisive Points
Lines of Operation and Lines of Effort	Lines of Operation and Lines of Effort
Direct and Indirect Approach	
Anticipation	
Operational Reach	Operational Reach
Culmination	Culmination
Arranging Operations	Phasing and Transitions
Forces and Functions	
	Basing
	Risk

Table 32: Joint Elements of Operational Design and Army Elements of Operational Art as listed in Joint Publication 5-0, *Joint Planning*, and Army Doctrine Reference Publication 5-0, *The Operations Process*, respectively.

———————————————

Termination criteria—The specified standards approved by the President and/or the Secretary of Defense that must be met before a military operation can be concluded. (JP 3-0)

—*DOD Dictionary of Military and Associated Terms*, 2018

The establishment of termination criteria provides a clear example of the interface of the military strategic with the grand strategic. The president or secretary of defense, often in dialogue with senior military leaders, will approve termination criteria. Once established, the joint force commander(s) will implement them. Because this activity happens at such high levels, the element itself is not a particularly useful concept for tactical planning, but because termination criteria link policy goals to military action and thereby drive further military planning, an understanding of them is essential. Until the 2020 version of Joint Publication 5-0, *Joint Planning*, joint doctrine listed termination criteria first because the other elements may be developed once this foundation is set, embodying the notion of beginning with the desired end in mind.*

Important as they are, the army does not consider termination criteria an element of operational art. This situation arises, perhaps, because army forces, as a matter of process, do not need to directly consider termination criteria in most cases. As mentioned earlier, joint force commanders implement termination criteria, meaning that they prepare for and execute combat operations. The services—army, navy, air force, marines, space force, and coast guard—provide forces to the joint force commander. The services themselves provide organizing, training, and equipping functions; they do not fight. Army forces provided to joint force commanders, then, would likely be more concerned with the military end state unless an army headquarters was serving as the nucleus of a joint headquarters.

Space operations forces, in contrast, can ill afford to ignore termination criteria. The "conditions that must exist in the [operating environment] at the cessation of military operations" necessarily involve the posture of space forces, the condition of mission infrastructure, and a consideration of the dependencies of the nation, including its fighting forces, on space capabilities.†

These considerations, and others, will likely tie directly to other policy objectives. In terrestrial war, termination criteria may include the reestablishment of international boundaries or legitimate governance, the reduction of an offensive threat to the region, the establishment of reliable internal

---

* The 2020 version of Joint Publication 5-0, *Joint Planning*, removed termination criteria as an element of operational design. We have chosen to leave the discussion here in case it is useful for any future audience. Doctrinal ideas, after all, have a tendency to rise from the dead. They are rather like acquisition programs in this regard.

† US Joint Staff, JP 5-0, *Joint Planning* 2017, IV-19.

security forces, and the posturing of combat power in theater to support the postwar environment.

A war that extends into space requires analogous considerations. The process of developing termination criteria, then, provides a means by which the military strategic and grand strategic coordinate action in reference to friendly space forces, enemy space forces, and civil space operations. If, for example, a policy objective of the conflict is the establishment—through diplomatic agreement—of new economic relations between the United States and the belligerent nation, one must consider the dependencies of the economy on the Global Positioning System (GPS). To that end, a necessary condition of the operating environment may be that the enemy no longer has the capability to degrade GPS operations. As long as the enemy retains the capability, the operating environment is not conditioned for conflict termination.

Furthermore, to formally conclude hostilities, friendly inspector satellites may have to visit a certain percentage of enemy satellites and verify their compliance with established norms, ensuring that they pose no threat to the GPS constellation (or to others). Such inspections may be precursors to negotiations covering antisatellite weapons, future prelaunch inspection/research facility inspections, or the return of control of captured or compromised space warfare systems (both friendly and commercial). With regard to the enemy force, possible termination criteria may include the capture or destruction of a sufficient portion of enemy space weapons systems, including launch and research facilities, to allow for long-term disarmament or force reduction goals; repositioning friendly systems to hold enemy systems at risk for leverage in diplomatic negotiations; the de-orbit of satellites to degrade intelligence capability; and the movement of captured orbital systems outside of predefined threat zones or the significant reduction of their on-board fuel reserves.

As with other elements, military planners can envision termination criteria for given scenarios in a preconflict environment. It is highly unlikely that a war will unfold as envisioned, but anticipating what may happen (anticipation is an element of operational design) allows the force to set the conditions in a theater (or multiple theaters) in advance of a conflict. Forces, time, and money are limited resources. However, it is possible to set conditions in such a way as to cover a multitude of contingencies, reducing overall risk for the most dangerous scenarios (more on this in Chapter 5). This opportunity is especially ripe for space operations forces, which operate globally and are less confined to traditional area-of-responsibility boundaries than an army or a fleet. Among the challenges in setting these conditions, however, are the long lead times and cost of fielding new satellites and ground segment equipment. By the time such systems are fielded, it is highly likely that the scenarios for which they were envisioned are no longer at the edge of the warfighting state of the art. As Douhet wrote in 1928, "the war in the air will be decided by those aerial forces which are in being and ready when hostilities break out," and the same may be said for space forces.* Secretary Rumsfeld's statement in 2004 was equally true, yet because he made the statement in the crisis environment of postinvasion Iraq, the notion that "You go to war with the army you have, not the army you might want or wish to have at a later time" was not so well received.† Indeed, a fundamental problem for which the space force was created was to address the historical inadequacy of space systems acquisition and fielding—an issue that touches the grand and military strategic as well as the institutional and warfighting aspects of space operations.

In a tactical sense, because of the difficulties of moving satellites to new orbits and operating theaters, when a space war breaks out, one can only employ the assets that are already placed in critical orbits and probably cannot rely on allies to make dramatic movements of their own space assets. These major movements of satellites take too much time and too much fuel, significantly shortening their useful lives and rendering them operationally useless during movement. So, for a war in the Pacific, for example, NATO space systems probably would not be much help, especially if space engagements conclude rapidly. The termination criteria of such a conflict,

---

*   Giulio Douhet, *The Command of the Air, in Roots of Strategy* (Book 4), ed. David Jablonsky (Mechanicsburg, PA: Stackpole Books, 1999), 396.

†   Eric Schmitt, "Iraq-Bound Troops Confront Rumsfeld over Lack of Armor," *New York Times*, December 8, 2004, accessed September 20, 2020, https://www.nytimes.com/2004/12/08/international/middleeast/iraqbound-troops-confrontrumsfeld-over-lack-of.html.

however, could extend to activities outside the Pacific, even to the point of strengthening NATO's space warfare advantages.

End state—The set of required conditions that defines achievement of the commander's objectives. (JP 3-0)
—*DOD Dictionary of Military and Associated Terms*, 2018

End state (Army)—A set of required conditions that, when achieved, attain the aims set for the campaign or operation. (See also battlefield visualization, commander's intent, and operation order (OPORD).) See FMs 71-100, 100-5, and 100-15.
—FM 101-5-1, *Operational Terms and Graphics*, 1997

Commanders at all echelons bear responsibility for determining desired military end states. For the joint force, these specifically military criteria enable the termination criteria as approved by the secretary of defense or the president. The army describes the complementary concept of "end state and conditions," which are typically defined in relation to friendly forces, enemy forces, and the civilian population. To continue the earlier example, if a condition of the operating environment at termination is that the enemy no longer has the capability to degrade GPS operations, part of the commander's end state may be the establishment of a local positioning, navigation, and timing network to offset remaining interference, the destruction of a certain number of enemy electronic warfare units, and the capture of enemy counterspace research facilities. These broad aims sound straightforward, but the challenge lies in the transition from conceptual planning (i.e., the listing of broad aims) to detailed planning. The detailed plan must translate the military end state into achievable tasks and objectives for one or more subordinate units to accomplish.

It is interesting that the older definition of end state (from 1997, epigraph) ties the military end state not to the achievement of objectives but to the attainment of aims. The word *aims* implies a more flexible, intent-based operation, perhaps one in which the campaign may attain its ends—even without the achievement of certain objectives. Indeed, as the military situation evolves, some objectives designated in the formal plan may become irrelevant. It is quite possible that an overfocus on achieving a tactical objective may lead to a lost opportunity for operational or strategic gain. Furthermore, while it is desirable to have clearly defined end states, especially for tactical units, it is sometimes helpful to think in terms of *future states* rather than end states. The *Defense Space Strategy Summary* discussed in Chapter 1, for example, employed "desired conditions," rather than end states. While it is certainly true that conflict terminates and forces achieve end states, it is equally true that the end of one operation transitions into the beginning of another. Particularly at the strategic level, as Dolman states, military activity is always seeking to gain renewed advantage, is always in the process of becoming.* To that end, it is essential that commanders understand the purpose of the objectives they are seeking so that they can take advantage and aid in achieving larger aims if given the opportunity.

In their pursuit of the British Army through Burma in 1942, the Japanese Army missed just such an opportunity. During the British withdrawal from Rangoon, the Japanese established a blocking position on the Prome Road—a line of communication on which the mechanized British greatly depended. As Slim recounted, the Japanese commander achieved his tactical objective of blocking the road, which allowed the Japanese main body to proceed to Rangoon as their campaign objectives required. That done, he removed the block, thus allowing British forces to escape to the north and sacrificing the opportunity for the destruction of the British Army in Burma.† This example highlights the need for even the tactical commander to understand the larger picture, to execute disciplined initiative when opportunities arise, and to build flexibility into plans to allow for such occurrences. These goals may prove particularly challenging in space operations because of the narrow focus of many mission areas and the constraints that surround operational decision making, particularly

---

*   Everett Dolman, *Pure Strategy* (New York: Frank Cass, 2005), 130–131.

†   William Slim, *Defeat into Victory* (New York: Cooper Square Press, 2000), 15.

with political sensitivities of space weapon employment. With proper training, clearly delineated authorities, and in-depth understanding of the plan, however, these goals are not insurmountable.

Objective—1. The clearly defined, decisive, and attainable goal toward which an operation is directed. 2. The specific goal of the action taken which is essential to the commander's plan. See also **target.** (JP 5-0)
—*DOD Dictionary of Military and Associated Terms*, 2018

Objective—1. The clearly defined, decisive, and attainable goal toward which every operation is directed. (JP 5-0) See ADRP 5-0, ATP 3-06.20. 2. The specific target of the action taken which is essential to the commander's plan. See ATP 3-06.20. 3. (Army) A location on the ground used to orient operations, phase operations, facilitate changes of direction, and provide for unity of effort. (ADRP 3-90)
—ADRP 1-02, Terms and Military Symbols, 2016

Objective (JP 1-02, NATO)—The physical object of the action taken, e.g., a definite tactical feature, the seizure and/or holding of which is essential to the commander's plan. (Army)—1. The physical object of the action taken (for example, a definite terrain feature, the seizure or holding of which is essential to the commander's plan, or, the destruction of an enemy force without regard to terrain features). 2. The clearly defined, decisive, and attainable aims which every military operation should be directed towards. 3. The most important decisive points. (See also decisive point.) See FMs 1-111, 6-20, 7-20, 7-30, 17-95, 71-100, 71-123, 100-15, and 101-5.
—FM 101-5-1, Operational Terms and Graphics, 1997

The objective is the only element of operational design that doubles as a principle of war, perhaps justifying its omission on the list of elements of operational art. Indeed, from Jomini to Clausewitz, from Mahan to Fuller, military theorists emphasize the concept, and practitioners employ it.

The objective is perhaps what the layperson thinks of when considering military operations: securing a bridgehead, destroying an enemy, seizing a hill. Objectives relate directly to the desired end state, and just as each military commander determines her own end states (as nested within those of the higher command's), each command is responsible for designating its own objectives to meet that end state. In so doing, each subordinate commander contributes to the attainment of the termination criteria or "the end of the war."

A common confusion among military practitioners is the difference between objectives and decisive points (see the following discussion on decisive points). It does not help that the definitions of *objective* include the word *decisive* in them—just as the definition of *end state* contains the word *objective*. The definition from 1997 goes so far to say that objectives are the most important decisive points, an association that has since been removed from the doctrinal definitions. So, what is the difference?

Importantly, objectives contribute directly to the overall achievement of the end state. Decisive points may be thought of as keys to the objective. They are points in time or space that provide a "marked advantage." The attainment of a tactical objective—the seizure of the satellite control ground station, for example—certainly provides an advantage to everyone involved, but the actual seizure is a decisive point for the operational-level commander, not the tactical commander. In all likelihood, the tactical decisive point occurred before the actual seizure of the objective, at the point when combat power and initiative were sufficient to ensure (as completely as possible) the attainment of the objective. If the ground station was lightly guarded, the tactical decisive point may have been when two platoons of troops had successfully disembarked from their helicopters at the landing zone. This action did not achieve the objective, but in so achieving a mass of their forces, the friendly force attained a marked advantage—a decisive point. This conception works because of the nesting of end states, objectives, and decisive points among superior and subordinate units. The higher unit's desired end states lead to objectives for its subordinates. In achieving their objectives, the subordinate units help their higher headquarters attain its decisive points.

One subordinate unit, designated as the main effort, will likely be the one to actually accomplish the objective on behalf of the superior headquarters.

If one component of the termination criteria in our fictitious scenario is assured access to GPS, destroying the enemy's GPS jammers is a direct-approach solution to this problem (see the discussion on direct and indirect approaches in the following section). The commander discussed in the section on end states already has this mission. However, as necessary as destroying enemy GPS jammers may be, the operational-level end state may also require reduction of the enemy's ground-based SDA radar network in a region. GPS and SDA are often treated as two separate missions, but indirectly, a degraded SDA network would limit the enemy's ability to track and target GPS (and other) satellites.* To that end, the objectives of a supporting commander might include the seizure of a command and control node by ground forces, the destruction of a radar site by cruise missile attack (an objective that equates to a target), the capture of another radar to be used for reverse engineering, the negation of another ground station through cyber effects, and the negation of orbital assets through a coordinated, multidomain approach. Denying the enemy's ability to observe terrestrial operating areas or disrupting their ability to command and control their space forces may serve as a supporting objective.

Effect—1. The physical or behavioral state of a system that results from an action, a set of actions, or another effect. 2. The result, outcome, or consequence of an action. 3. A change to a condition, behavior, or degree of freedom. (JP 3-0)
—*DOD Dictionary of Military and Associated Terms*, 2018

---

* One might rightly argue that targeting could still occur because GPS satellites follow regular orbits and will therefore occur predictably. This is true enough, and once the SDA network is degraded, orbital readjustments would have to happen to prevent such predictability.

Effect—(DOD) 1. The physical or behavioral state of a system that results from an action, a set of actions, or another effect. 2. The result, outcome, or consequence of an action. 3. A change to a condition, behavior, or degree of freedom. (JP 3-0) See FM 3-53, FM 3-57, ATP 3-09.24, ATP 3-53.2, ATP 3-57.60, ATP 3-57.70, ATP 3 57.80, ATP 3-60.
—ADRP 1-02, Terms and Military Symbols

In the case of *effects*, the joint and army definitions are identical. The interesting difference between the preceding pair of definitions is that the army lists a number of doctrinal references that the joint definition does not, providing important clues about the concept of effects. The two listed field manuals are FM 3-53, *Military Information Support Operations* and FM 3-57, *Civil Affairs Operations*. The Army Techniques publications are (in order as listed) for the fires brigade, information in conventional operations, civil affairs planning, the civil-military operations center, civil-military engagement, and targeting. What is also telling is that the army's 1997 lexicon, usually more thorough than the 2016 version, does not include *effects* as a term.

The term *effects* is a reflection of the evolution of military thought during the Global War on Terror period. Its emphasis on a systems approach, the incorporation of information operations, and the nonkinetic aspects of fires (for example, electromagnetic spectrum operations) was well suited to counterinsurgency operations and will continue to be indispensable for future military operations. As important as the concept is, however, the term has been overused. The idea of "effects-based operations" that came into vogue during the Iraq and Afghan Wars, ran the risk of reducing operations to a concept of a single way—effects-based targeting—albeit with multiple means of prosecuting those targets. In space operations, the discussion of effects continues to dominate professional discourse, nearly to the exclusion of the entirety of the rest of the warfighting vocabulary.

The lessons learned regarding information and nonkinetic operations should not be discounted in the future, but the space force would do well to remember that *effects* is only one of many concepts, akin in its functionality to the ideas of both *purpose* and *consequence* (not solely one or the other, as the term is often used). At the operational level, one arrays effects against an enemy center of gravity. A desired effect, like a purpose, is the reason for conducting a military action. The assessed effect, however, may, because of the complexities of the system, result in a different outcome (i.e., a consequence) than the one desired.

As a second word of caution, the effects-based approach has led to a significant amount of negative training over the years.* During exercises, it is not uncommon for a commander to ask for space or cyber effects (or for an overeager staff officer to promise such). By speaking of effects, such staff officers could sidestep the tricky problem of classified means. Essentially, because of classification concerns, the supported commander could not know *how* the space professional was contributing; he could only know *what* the space professional was contributing and extrapolate that into the *so what* and *therefore* of the operation. Further compounding the problem, most military exercises employ underdeveloped space scenarios, focusing instead on the land, air, and maritime components. In this environment, space and cyber become an afterthought, commanders request effects, and because it's an exercise, the effects automatically happen. In real life, however, the employment of space and cyber techniques is far from automatic.

On the contrary, a significant amount of intelligence analysis and operational preparation must occur, which is why it is essential for experts in these fields to be involved early in planning for both exercises and for real-world contingencies. Detailed and stressing scenarios should include fully developed courses of action aimed at testing the flexibility of the friendly force plan, possible enemy counteractions, any battle management systems, and the ability of the commander to incorporate space warfare into his decision cycle. In addition, a smart adversary would employ some of his space systems in a manner different than in peacetime to confuse historical intelligence collection and maybe even implement war-reserve modes that have not been detected before. Furthermore, exercises present an opportunity to test new tactics, techniques, and procedures; to explore

---

*   The authors would like to thank Lt. Col. Brandon "Coach" Davenport, US Space Force, for bringing this notion to our attention.

the acceptability and consequences of system loss, unit loss, and collateral damage; and to probe the boundaries of authorities and interests among military departments, government agencies, and allies. In short, exercises allow for the assessment of the many unknowns associated with space warfare and gaining confidence before a crisis occurs.

Center of gravity—The source of power that provides moral or physical strength, freedom of action, or will to act. Also called **COG**. See also **decisive point.** (JP 5-0)

—*DOD Dictionary of Military and Associated Terms*, 2018

Center of gravity—(DOD) The source of power that provides moral or physical strength, freedom of action, or will to act. Also called COG. (JP 5-0) See ADRP 3-0, FM 3-24, ATP 305.20, ATP 3-53.2, ATP 3-57.60, ATP 3-57.80, ATP 3-92, ATP 5-0.1.

—ADRP 1-02, Terms and Military Symbols, 2018

Centers of gravity (JP 1-02)—Those characteristics, capabilities, or localities from which a military force derives its freedom of action, physical strength, or will to fight. (Army)—The hub of all power and movement, on which everything depends. (See also operational art and operational level of war.) See FMs 100-5, 100-7, and 100-15.

—FM 101-5-1, Operational Terms and Graphics, 1997

With centers of gravity, we again have a case in which the contemporary definitions mirror one another. Again, the army's list of reference publications hints at the widespread adoption of the concept in its operations and counterinsurgency field manuals and in its techniques publications for special operations intelligence, information in conventional operations, civil affairs planning, civil-military engagement, corps operations, and design methodology. Despite the ubiquity of the concept in contemporary doctrine, the definitions suggest the singularity of the concept—the notion that there can be only one center of gravity, and that center is a Clausewitzian source of moral or physical power. The army definition of 1997 follows this rationale, but the joint definition of 1997 is more expansive, allowing for multiple potential centers of gravity.

Indeed, in a given military operation, centers of gravity may exist in every adversarial pairing. In like pairings, a fleet may seek the center of gravity of the enemy fleet, a numbered air force of a numbered air force, a field army of a field army. These pairings, while practical, are also predictable and less creative than solutions involving nonlike pairings. For example, it is highly possible that a cyber force could attack the center of gravity of a space force (or objectives related to the center of gravity) or that an air force or ground force could do the same. The employment of all joint warfighting options (and if available, combined warfighting options) will help ensure that the enemy faces more dilemmas than they can effectively handle, especially if employed in a coordinated and time-sequenced manner against tactical, operational-level, and strategic centers of gravity.

The designation of a center (or multiple centers) of gravity is highly dependent on the situation, and despite routine efforts to apply an objective process, the determination of the center of gravity for a given scenario remains subjective, a case for planners' and commanders' judgment. To its credit, air force doctrine invokes this sense of uncertainty and the danger of thinking of the enemy as a static being with static centers of gravity and offers several analytical approaches to arriving at a determination in the matter.* In a space warfare scenario, one may designate a GEO belt sector, the ability to disseminate information, political or popular will, or even the coalition as a center of gravity (as General Schwarzkopf did before the Persian Gulf War).† Similarly, if one applies the Strange method, one may designate any number of centers of gravity and assess critical requirements,

---

*   Appendix 1 to Annex 3-0 offers the Joint COG model (also known as the Strange model after Dr. Joe Strange, which originated within the Marine Corps), the Strategic Ring COG model (an air force outgrowth), the National Elements of Value model, and the "CARVER" method as potential viable methods. US Air Force, Annex 3-0 Operations and Planning, Appendix A: Center of Gravity Analysis Methods, November 4, 2016, accessed June 1, 2020, //www.doctrine. af.mil/Portals/61/documents/Annex_3-0/3-0-D30-Appendix-1-COGAnalysis.pdf.

†   US Joint Staff, JP 5-0, *Joint Planning* 2017, IV-26.

capabilities, and vulnerabilities from there, gaining insight into the enemy's systems. Alternatively, in an application of Warden's Five-Ring model to space warfare, the rings may be thought of as space leadership, command and control communications, space domain awareness infrastructure, scientists and technicians, and space forces, weapons, and launch facilities (see Figure 76). Application of such methods can be very useful to gain insight about the enemy (or about one's own forces, if applied in reverse), but they must be applied with great critical thought as part of the planning process—not as a sanctioned method for arriving at the "correct" answer.

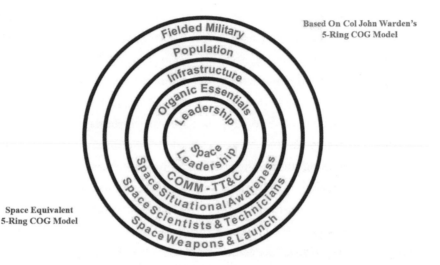

Figure 76: A Mapping of Warden's Five-Ring Model to Space Operations. *Source:* Originally published by Paul Szymanski, "Techniques for Great Power Space War," *Strategic Studies Quarterly*, vol. 13, no. 4 (November 2019).

Whatever the method, it is important to note that the successful action against a center of gravity does not in itself guarantee victory—assuming the analysis did not lead to a false conclusion in the first place. A thorough analysis may suggest as centers of gravity command and control facilities that support multiple missions, regions of space that house critical capabilities, or the political will of belligerents to employ their space systems with confidence. At the operational level, a center of gravity is most often an enemy force. In any case, even if one center of gravity is destroyed, alternative means may be available to make up for its loss until such time that the enemy can reconstitute it. Reducing the vulnerability of friendly force centers of gravity to enemy attack through enhanced protection or deception measures, through mobility, or even through training becomes a problem for all practitioners of warfare and, in the long term, for the institution itself. It is thus critical to perform a center of gravity analysis not only on the enemy force but on one's own force, as well.

***

Decisive point—A geographic place, specific key event, critical factor, or function that, when acted upon, allows commanders to gain a marked advantage over an enemy or contribute materially to achieving success. See also **center of gravity.** (JP 5-0)
—*DOD Dictionary of Military and Associated Terms*, 2018

The preceding section on objectives discussed the concept of decisive points and their relation to objectives. In a sense, decisive points are actually more important to mission success than objectives because decisive points are the gateway to achieving objectives. If one is employing the principle of mass, one masses forces at the decisive point—not necessarily at the objective. The decisive point unlocks the objective; the objective weakens or destroys the center of gravity; without the center of gravity, the enemy, in theory, cannot prevent the friendly force from achieving its end state. Determining the decisive point thus allows for a more refined understanding of the aspect of the objective that is the most critical. Here the commander masses his combat power, including, as J. F. C. Fuller would say, mental, moral, and physical force.*

---

*   Combat power, of course, has its doctrinal definitions, which vary between the Joint Force and the army. Combat power—(DOD) The total means of destruction and/or disruptive force that a military unit/formation can apply against the opponent at a given time (JP 3-0), see FM 3-07, ATP 3-90.5; (Army) The total means of destructive, constructive, and information capabilities that a military unit or formation can apply at a given time (ADRP 3-0); Fuller, *Foundations of the Science of War*, 202–207.

For Jomini, the concept of *points* expanded beyond the decisive point and included three subtypes: the geographic decisive point, strategic points of maneuver, and the decisive strategic point. As an example of a geographic decisive point, he mentions the city of Leipzig as "the junction of all the communications of Northern Germany."* Applying our modern lexicon, we may consider the seizure of Leipzig a decisive point of the campaign (in the military-strategic sense), or Leipzig itself may actually be a terrain-based objective for a portion of the force to seize—an interpretation in keeping with the army's historical conception of objectives as the most important decisive points.

Next, Jomini uses the term *strategic points of maneuver*, which have value because of their relation to the enemy.† As a rule, he says, maneuvering to the flank of an enemy is desirable, but more generally, Jomini is discussing the idea of arranging operations (another element dealt with subsequently) in such a way that puts friendly forces into a position of relative advantage. Prior to U. S. Grant's siege of Vicksburg in the summer of 1863, he met Pemberton's forces at the Battle of Champion Hill. In Jominian terms, he had positioned his army at a strategic point of maneuver, and if he could destroy the Confederates outside the walls of the city (the objective), a costly siege would not be necessary. With most of the local Confederate Army engaged, Grant likely recognized this engagement as both the operational-level decisive point of the campaign and also as a strategic decisive point in the war.

For Jomini, either the decisive point or the strategic point of maneuver could be classified as a *decisive strategic point*, if "its importance is constant and immense."‡ These points are the ones that are "capable of exercising a marked influence either upon the result of the campaign or upon a single enterprise" and are most akin to the concept behind the doctrinal decisive point.§ They are different in degree from other decisive points, are of strategic consequence, and may become objectives themselves. In Grant's case,

---

*   J. D. Hittle, ed., *Jomini's Art of War* (Harrisburg: Stackpole, 1965), 73.

†   Ibid.

‡   Ibid.

§   Ibid.

Vicksburg itself was a geographic decisive point because it controlled the Mississippi River. In the course of the campaign, Grant attained a strategic point of maneuver vis-à-vis Pemberton. Both points rose to the distinction of decisive strategic points because either the destruction of Pemberton's Army in the field or the capture of Vicksburg would have contributed to the dual strategic aims of opening the Mississippi River Valley for Union control and bisecting the Confederacy into nonsupportable eastern and western halves.

Although the Union Army defeated the Confederate Army outside the city, Pemberton's army was not destroyed. It retreated inside Vicksburg's prepared defenses, thus making a costly siege necessary.

For space operations, we may imagine analogous cases. In order to provide continuous satellite communications support to a flotilla transiting the Atlantic Ocean, space forces would have to occupy and control a handful of orbital slots, the attainment of which would ensure communications among the homeland, the flotilla, European Command, and NATO. Like Leipzig, these orbital slots allow for the confluence of communications and, in Jomini's terminology, are geographic decisive points.

The vignette preceding this chapter provides an example of strategic points of maneuver. Assuming that China deliberately positioned its BeiDou satellites within striking distance of GPS satellites, we may say—figuratively, of course—that the BeiDou constellation was on the flanks of the GPS constellation, prepared to exploit its vulnerabilities. Like Pemberton's Army, the GPS constellation was exposed, and its destruction would have accomplished strategic objectives without the cost of protracted engagement. Because of its "constant and immense importance," the strategic point of maneuver, from China's perspective, may also be considered a decisive strategic point.

Line of operation—A line that defines the interior or exterior orientation of the force in relation to the enemy or that connects actions on nodes and/ or decisive points related in time and space to an objective(s). Also called **LOO.** (JP 5-0)
—*DOD Dictionary of Military and Associated Terms*, 2018

Line of effort—In the context of planning, using the purpose (cause and effect) to focus efforts toward establishing operational and strategic conditions by linking multiple tasks and missions. Also called **LOE.** (JP 5-0)
—*DOD Dictionary of Military and Associated Terms*, 2018

Of all the definitions of operational art and design terms, the definition of a line of operations (LOO) is, given the straightforwardness of the concept, perhaps the most confusing. In clearest terms, a line of operations is a route on the map that connects decisive points and leads to the attainment of objectives. This definition works well for the theater-level operations because it is in keeping with Jomini's conception. In Jomini's conception, "the object of the campaign determines the objective point," which is a singular outgrowth of strategy.* Other points—decisive geographic points, strategic points of maneuver, or decisive strategic points—contribute to this singular objective point. If we agree, however, that some decisive points are very near their objectives—indeed, will likely inform the development of objectives for subordinate units—then a LOO can also link a series of intermediate objectives that lead a force to its ultimate objective or to the overall end state. Strictly speaking, the definition seems to disallow this, but the question again turns to the interpretability of Jomini's work as his work has come down to us through the multiple filters of three centuries. Contemporary thought allows for the possibility of tactical, operational-level, or strategic objectives just as it allows for decisive points at these levels.

Joint doctrine explains an interior orientation as one in which LOOs of forces diverge as if from a central position. An exterior orientation is one in

---

* Hittle, 74.

which LOOs of forces converge (as in an envelopment).* Interior and exterior LOOs are a Jominian concept, as well, but the preceding definitions lack intuitiveness, even when explained by Jomini.† A more intuitive explanation of internal and external orientation LOOs comes from a consideration of lines of communication (LOCs), which is neither an element of operational art or design but is the single most important theoretical concept in space operations (more on that later). If lines of operation extend forward, lines of communication extend rearward, connecting the fighting force to their base(s). In a central position—the Union Army at Gettysburg—the lines of communication are *interior* to the fighting position, allowing rapid reinforcement; the LOOs are said to have an interior orientation. Conversely, a position that is encircling a central position—the Confederate Army at Gettysburg—has lines of communication that radiate rearward from the front (exterior). In this case, moving reinforcements and supplies is more difficult because the lines do not connect to a centralized base and because they have further distances to travel as they redistribute forces along the circumference of the battle position.

Lines of effort (LOEs) similarly link decisive points, but rather than linking them together in physical space and time as LOOs do, LOEs link decisive points in conceptual space and time. LOEs can prove a useful organizing principle and were common in the stability operations of the Iraq and Afghan Wars. They are also particularly useful when dealing with institutional or organizational planning (as demonstrated by the *Defense Space Strategy Summary*). A common oversight in such applications, however, is the visualization that shows straight arrows for each line of effort, leading to a defined end state to which each effort apparently arrives at the same time. Such visualizations show a lack of critical thought because they often do not depict the interdependencies of the lines of effort or the sequencing of decisive points or objectives. The LOE chart offered in JP 5-0 provides an example of the former, static approach to LOE employment of the kind that links thematic elements of a counterinsurgency operation (Figure 77).

---

*   US Joint Staff, JP 5-0, *Joint Planning* 2017, IV-28.

†   Hittle, 78.

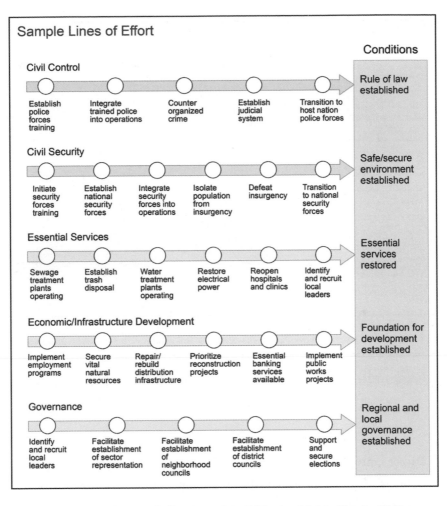

Figure 77: Sample Lines of Effort, *Source:* Joint Publication 5-0, *Joint Planning*, IV-30.

In contrast, the army's strategic approach from 2018 provides an example of LOE employment that is more in keeping with the spirit of the line of operation (Figure 78). This illustration highlights the fact that, even though LOEs exist in conceptual space, they exist in this space relative to the other LOEs, to the anticipated threat landscape, and to time. Furthermore, the following graphic goes so far as to designate main efforts, label objectives, and depict how those contribute to the attainment of the desired future state.

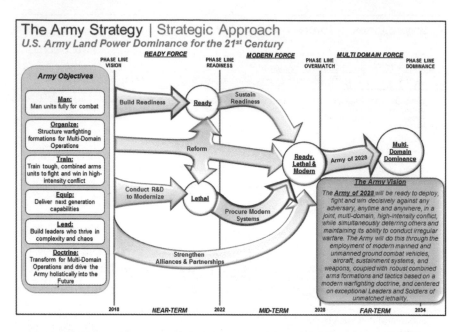

Figure 78: The Army's Strategic Approach from 2018. This graphic demonstrates an integrated approach as visually depicted by interdependent LOEs that converge and diverge in conceptual space. *Source:* US Department of the Army, "The Army Strategy," n.d., accessed September 20, 2020, https://ww" https://www.army.mil/e2/downloads/rv7/the_army_strategy_2018.pdf.

Direct approach—The manner in which a commander attacks the enemy's center of gravity or principal strength by applying combat power directly against it. (ADRP 3-90)

—ADRP 1-02, Terms and Military Symbols, 2018

Indirect approach—The manner in which a commander attacks the enemy's center of gravity by applying combat power against a series of decisive points while avoiding enemy strength. (ADRP 3-90)

—ADRP 1-02, Terms and Military Symbols, 2018

Like LOOs and LOEs, the indirect and the direct approach are related concepts. Even though they are joint terms, neither is defined in the DOD dictionary, so we must rely on army definitions, which reflect the joint doctrinal explanation. In addition to the doctrinal definition, the joint explanation of the indirect approach adds that an "indirect approach attacks the enemy's COG by applying combat power against critical vulnerabilities that lead to the defeat of the COG while avoiding enemy strength."* Whether decisive points as stated in the army definition or critical vulnerabilities as stated in the joint explanation, perhaps both are too analytical, the joint explanation forcing a center of gravity and critical vulnerabilities analysis à la the Strange method. The theoretical concept is more nuanced than doctrine suggests.

The British theorist and military historian Sir Basil Liddell Hart explains the theoretical concept of the indirect approach in his book *Strategy*, and the direct approach only emerges in its afterglow. In Liddell Hart's estimation, history's most successful commanders succeeded not by attacking the enemy's strongest points in the most expected ways but by employing forces in unexpected ways that left more orthodox commanders at a loss. Their "decisive points" often resided within the moral sphere; a realization that makes the clinical doctrinal definitions uncomfortable. In *Strategy*, Liddell Hart offers numerous historical examples of the success of the indirect approach from Belisarius to Sherman, yet perhaps the starkest example comes from Sir Basil's contemporary T. E. Lawrence.

If the Western theater of World War I championed the direct approach—massed forces against the enemy's strong points—the Eastern theater, under the constraints of economy of force, allowed for an indirect approach. It was under these conditions that Lawrence and Faisal bin Hussein created an Arab insurrection against the Ottomans in Arabia. Relying on guerrilla tactics, the Arab forces harassed the Turks in that theater until General Allenby was able to mount his advance into Palestine and ultimately Syria. If the British regulars of the campaign largely represented the direct approach, Lawrence's complement to it represented the indirect approach. In the example of Lawrence, space operators may find a particularly useful, if unexpected, champion. In Arabia, Lawrence acknowledged that it was not necessary to occupy or even control the whole of the Hejaz; the Arabs needed only to be

---

*   US Joint Staff, JP 5-0, *Joint Planning* 2017, IV-33.

in the right places with the ability to seem like they were everywhere. It was not worth wasting resources on occupying territory or securing objectives that did not truly matter.* Certainly one cannot control all of outer space, particularly if one starts to consider defense of translunar space along with near-earth orbits, but it may be possible to control enough of space to allow Allenby to reach Damascus. For this reason, fake and decoy space threats may be just as effective as real antisatellites for deterrence purposes.

***

Anticipation—The ability to foresee operational requirements and initiate actions that satisfy a response without waiting for an operation order or fragmentary order. (ADP 4-0)
—ADRP 1-02, Terms and Military Symbols

The entire planning process revolves around anticipation—guessing about what the enemy may do, foreseeing how friendly forces may stumble or fail, attempting to posture the force in such a way that it can react to the unknowable. The first responsibility belongs primarily to the intelligence arm, the second to the operations arm, the third with all arms. In light of this, it is interesting that the definition offered by the army cites only sustainment doctrine as a reference (Army Doctrine Publication 4-0). Anticipation is the reaction to the fear of the principle of surprise; it is how commanders and staffs feed their imaginations forward and, when they return, develop responses to outwit them.

The problem of anticipation for space operations is a particularly thorny one. The environment, including the orbital and electromagnetic environments, is external to the human experience in a way that land, air, and sea operations are not. Activities within those environments are not easily visualized (especially with the vast distances involved preventing direct visualizations of satellite threats and electromagnetic activity), and enemy activities or responses may not have historical precedent. What's more, the

---

*   T. E. Lawrence, *Seven Pillars of Wisdom*, Chapter XXXIII.

space operations problem set is a global one in which a relatively small set of cadre need to be as expert as possible in the diverse missions, threats, and plans.

The primacy of space domain awareness (of the orbital, link, and ground segments) again raises its head. After all, without sensor information about what is happening, one cannot anticipate the next move or the move-after-next. But anticipation is more than just a technical problem. It is a concern of the institution to develop creative thinkers with a sound basis in the fundamental language of warfighting, thinkers who are able to interact with other domain plans in an intimate way and communicate their perspectives effectively.

Operational reach—The distance and duration across which a force can successfully employ military capabilities. (JP 3-0)
—*DOD Dictionary of Military and Associated Terms*, 2018

Any force, no matter how well trained or resourced, operates on a finite amount of resources. Without rest and resupply, the force can only reach so far. If it encounters enemy resistance, poor weather conditions, or suffers from low morale, its operational reach shortens even further. From a planning perspective, the objectives are to determine what the realistic operational reach of the force is, how to extend it, and how to sequence units so that the entire force does not reach the end of its tether at the same time.

Space operations forces encounter operational reach in a variety of flavors. On orbit, satellites are limited in their movement by finite power and fuel and by effective ranges for most weapon systems (even laser weapons). Low earth orbit satellites may only be overhead and in communications with a specific ground controller for five to eight minutes, a temporal component of operational reach. At other times, operators may not be aware of what an adversary may be doing against a satellite when it is out of the reach of sensors. Figure 79 shows a geographic representation of when some satellites may not be available to ground controllers (red portions depict a maximum time delay to next access), causing some parts of a satellite's orbit to be more vulnerable to attack than others.

In the electromagnetic spectrum, geography limits operational reach. A command-and-control ground station in the Middle East, for example, cannot directly uplink commands to a satellite flying over Mexico. On the ground, however, where space operations begin and end, resides the most significant limitation to operational reach—the human operator. Because of this factor, the operational reach for space operations is less limited by physical distance than by duration. Space operations are continuous and require constant engagement even in a peacetime orientation. In a wartime orientation, energy will fail even the most motivated and professional troops, limiting the ability of the force to fight on.

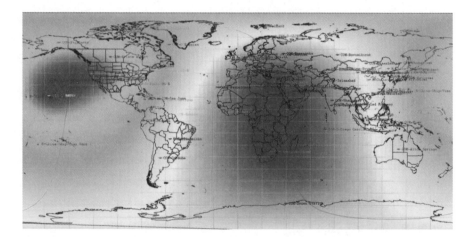

Figure 79: Example Attack Orbital Locations Optimized for Allied Space Surveillance (Blue) and to Avoid Adversary Space Surveillance Sensors (Red). *Source:* Paul Szymanski, "Issues with the Integration of Space and Terrestrial Military Operations," *Wild Blue Yonder Online Journal* (June 2020). https://www.airuniversity.af.edu/Wild-Blue-Yonder/Article-Display/Article/2226268/issues-with-the-integration-of-space-and-terrestrial-military-operations.

Culminating point—(Army) That point in time and space at which a force no longer possesses the capability to continue its current form of operations. (ADRP 3-0)
—ADRP 1-02, Terms and Military Symbols, 2016

Culminating point—The point at which a force no longer has the capability to continue its form of operations, offense or defense. (JP 5-0)
—DOD *Dictionary of Military and Associated Terms*, 2018

Culminating point—The point in time and space when the attacker can no longer accomplish his purpose, or when the defender no longer has the ability to accomplish his purpose. This can be due to factors such as combat power remaining, logistic support, weather, morale, and fatigue. See FM 101-5.
—FM 101-5-1, Operational Terms and Graphics, 1997

Conceptually, the culminating point exists at the end of a unit's operational reach, but it would be wise for any force to leave some margin in this matter. If a unit is on the offense, for example, it will reach a point where it is no longer able to continue on the offensive and must transition to the defense. Such a transition, however, is easier said than done, and in fact, transitions themselves may be the most difficult part of any operation. An armored corps, for example, cannot simply stop forward movement and proclaim that it is now on the defense. It must employ its reconnaissance elements, find and construct suitable defensive positions, rework its fires plan, cross-level and bring forward supplies, establish medical facilities, and so on. The ability to culminate *and transition* is what distinguishes culmination from mission failure. Without the ability to anticipate and affect a transition, the culmination point marks the point in time and space where a force is ineffective and therefore highly vulnerable.

Culmination exists at all levels of war and in all domains. A master planner would consider not just the culmination of the armored corps but also the culminations of all the corps within the land force and the culmination of the land force with respect to air, maritime, space, and cyber forces. The fate of one is inextricably intertwined with the fates of the others, particularly in the highly networked operating environments of today. Managing these transitions will be a key consideration of any joint force commander in the future, but what exactly does it mean for space forces to culminate and transition?

Arguably, US space forces, such as they have existed historically, have remained on the strategic defensive. The nation has placed satellites into orbit in defensive postures to perform defensive missions of strategic significance. The first air force satellite program developed three capabilities on behalf of Strategic Air Command—missile warning, signals intelligence, and imagery intelligence—all directly related to protecting the United States from a Soviet missile attack. Other capabilities like satellite communications and positioning, navigation, and timing contributed to the palisade in space, and while they enabled offensive action in other domains, they were themselves not offensive in nature. To date, space operations have entered into the operational level for the defense with missions like missile warning, synchronizing the tactical activities of a host of units. Similarly, one may achieve operational-level activity by combining some tactical offensive actions with some tactical defensive actions. Available tactical examples, however, can hardly be said to transition from one to the other; their defensive or offensive activity simply ends.

With the exception of limited offensive potential at the end of the Cold War—ground-based lasers, on-orbit antisatellite weapons—the United States did not enter into other levels of space warfare. In fact, with the information available, it is difficult to argue that the United States has ever demonstrated a strategic-level or operational-level offensive in the space domain. The questions then become: what is a strategic-level offensive, what is an operational-level offensive, and what could those look like for space operations? Better yet, what do space operations look like as part of an all-domain operational-level or strategic-level offensive?

Based on this assessment, gaps exist in the ability of US space operations forces to engage in a strategic offensive and in an operational-level offensive or defensive. The catch, however, is that in space operations—as in operations in other domains—the tactical level cannot exist in direct contact with the strategic level. There must be an interface. Even a direct-strike tactical action with strategic consequences ordered by the president—think the special operations raid against Osama bin Laden—does not exist without the interface of operational art, that is, the synchronization of tactical action to meet strategic objectives. Similarly, the dropping of a nuclear bomb is a tactical action that may achieve a strategic objective, but this single action does not exist in a vacuum. It is supported by basing operations, air traffic controllers, intelligence collection, information operations, and aerial refueling, among others.

Arranging Operations

Arranging operations is not a defined term in joint or army terminology, but like anticipation, it is an intuitive concept. In some sense, the entire planning process is about the arrangement of operations as guided by considerations of termination criteria, military end state, objectives, and the other elements. As a concept, arranging operations includes arranging all types of operations, and this effort begins in the planning process. As a rule of thumb, it is helpful to classify the different types of operations by warfighting functions: command and control operations, intelligence operations, movement and maneuver operations, fires operations, protection operations, and sustainment operations. A fully developed course of action will include concepts for all of these operations. Of course, the concept of sustainment does not exist independently of the concept of movement and maneuver, for example. On a planning team, the movement and maneuver planners will coordinate with the sustainment planners to ensure that they are in agreement about the overall scheme. This is what is meant by having an integrated plan. The integrated plan of one echelon must also be integrated with the planning efforts of higher, lower, and lateral commands. Properly integrated plans allow for properly synchronized operations.

It is important to note that the word *operations* does not imply that such activity happens only at the operational level of warfare. In fact, arranging operations happens at all levels of warfare, and the arrangement of operations at the strategic level will drive considerations for arranging operations at subordinate levels.

Force—1. An aggregation of military personnel, weapon systems, equipment, and necessary support, or combination thereof. 2. A major subdivision of a fleet. (JP 1)
—DOD *Dictionary of Military and Associated Terms*, 2018

Function—The broad, general, and enduring role for which an organization is designed, equipped, and trained. (JP 1)
—DOD *Dictionary of Military and Associated Terms*, 2018

To plan for the employment of a military organization, a force, one must understand what the force is made of and what the force can do—its potential functions. Within a military career, depth of understanding of forces and functions is accrued over time through experience and professional military education, most of which are service-oriented. A new lieutenant of armor will learn his craft, focusing on the basics of leadership, the employment of a four-tank platoon, and the maintenance necessary to keep the platoon going. In so doing, his skill set will begin to broaden. He will begin to learn how transportation and supply systems work. He will coordinate with and learn about the employment of artillery, infantry, and engineers. All the while, he will be learning about the processes that keep the army rolling along.

Depending on his career experiences, he may have opportunities to work with air liaison or weather officers from the air force or with Air Naval Gunfire Liaison company officers (ANGLICOs) from the Marine Corps. As part of a joint or combined organization, he would begin to learn about the forces and functions available from the other services. As he becomes more senior, he learns about the institutions, perhaps as part of the army or joint staff in the Pentagon, and maybe he is even exposed to multinational partners through a staff college assignment with the Australians or through an assignment working with NATO. He may even be exposed to other instruments of national power through an assignment at, say, the

Department of Commerce. The point is that as one's career progresses, one is exposed to the different forces and functions of warfighting.

Figure 80 depicts this idea as a series of concentric ellipses. In the beginning, one learns one's specific trade within one's branch. These skills are decidedly tactical and focused (i.e., learning about tanks), but it is possible to learn things from other ellipses even while one is living inside another. The principle of combined arms comes along next. In colloquial usage, combined arms has come to mean the use of multiple forces together (tanks, infantry, artillery); in military theory, however, it actually means using the differing functions of multiple forces to create untenable dilemmas for the enemy, each of which is highly distasteful (more on dilemmas in the next section). The next ellipse is joint operations. In the warfighting sense, joint operations are like combined arms, but now you are dealing with the forces and functions of the other services.

Coalition operations add in forces and functions of Allied militaries, and the "national" ring includes whole-of-government activities, including defense agencies external to the joint force.

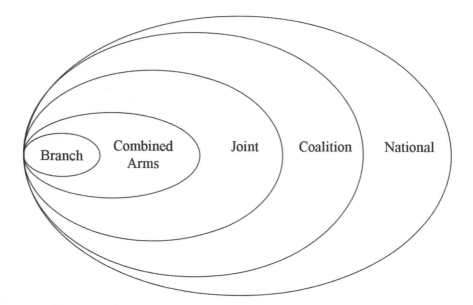

Figure 80: A Model for the Progression of Professional Expertise. *Source:* Created by Jerry Drew.

This model has significant implications for the space force moving forward. First, within the space community, only the space force and army have dedicated space officers. The marines and navy have additional duty space officers. For enlisted personnel and warrant officers, the army does not have a specific specialty but borrows from the air defense, communications, intelligence, and engineer communities. This, of course, is to say nothing of the host of other skills and specialties that are essential to a military force (sustainment, human resources support, cyber, electronic warfare, spiritual welfare). It may be tempting to propose space force branches along mission lines (SATCOM, PNT, ISR, and so on), but those specialties are perhaps too narrow.

Alternatively, the core branches of the space force—like the armor, infantry, and artillery of the army—could be terrestrial warfare, electromagnetic warfare, and orbital warfare. Company-grade officers would start in their primary branch, and as they progress, expand their knowledge into the other facets of the craft with the ultimate goal of having well-rounded field grade officers ready for joint assignments. While the officers seek breadth of experience, enlisted personnel would seek depth. The need for breadth of experience for officers will inhibit them from developing vast technical depth in any one mission. Similarly, enlisted personnel will become highly proficient at operating their systems but may not have the technical background to understand the inner workings of their equipment. For a combination of system experience, technical expertise, and leadership, the space force should establish a warrant officer corps to specialize in the technical aspects of each mission area.

Base—(DOD) A locality from which operations are projected or supported. (JP 4-0) See ADRP 3-0, FM 3-14, FM 4-95, ATP 3-91, ATP 4-94.
—ADRP 1-02, Terms and Military Symbols, 2016

Given the importance of basing, it is somewhat surprising that the joint elements do not offer it as a main consideration. Not only are bases necessary domestically, but establishing new bases in austere environments and maintaining a forward presence at key locations are essential activities for an American way of war that depends on power projection. In Mahan's words, the question of basing falls under his "establishment of depots" to connect to the "home base," which in that context, serve as supply links on the naval line of communication.*

As with forces of other types, space operations forces require domestic basing, forward basing, and depots. Indeed, domestic bases already exist in the forms of fixed-site ground operations stations like Schriever Space Force Base for satellite control and Patrick Space Force Base for launch operations. These both rely on a network of out-stations for data relay, observation, and backup functions. In other words, the main base does not exist without a network of other smaller bases to support its functions. Forward bases exist, too, in countries like Germany and Japan where satellite communications facilities provide in-theater access. These bases, such as they exist today, do not exactly fulfill the depot function explained by Mahan. Yes, they provide basic supply and maintenance, but they are not stockpiles of extra satellites, ground segment equipment, or launch vehicles that may be employed in the event of a conflict. The long-sought goal of "operationally responsive" space operations would require such basing—either in theater like army prepositioned stock or near a convenient intertheater transport hub that could take advantage of nearby shipping and/or air transport.

In the near future, a second type of basing may become available—orbital basing. These will likely not be manned space stations or moon bases but strategically positioned orbital platforms. Such platforms—motherships, if you like—may distribute satellites to assess damage, gather intelligence, repair broken solar arrays, or replace old batteries and empty fuel tanks. The advantage of such a platform lies in its depot function. If it could be assembled in space—a possible feat, even with today's technology—independent launches could position the various components—a module of spare batteries, a module of new fuel tanks—without the complications of

---

*   A. T. Mahan, *The Influence of Sea Power upon History*, in *Roots of Strategy (Book 4)*, ed. David Jablonsky (Mechanicsburg, PA: Stackpole Books, 1999), 63.

building a robotic satellite for each activity. Advances in artificial intelligence may even make such activities semiroutine.

Risk—(DOD) Probability and severity of loss linked to hazards. (JP 5-0) See ATP 5-19.
—ADRP 1-02, Terms and Military Symbols, 2016

Commanders own risk decisions. What risk is acceptable, based on likelihood and consequence, varies widely from situation to situation. During periods of routine operations, risk tolerance is likely to be lower; during wartime operations, risk tolerance will necessarily be higher. For space operations, risk-to-mission concerns have long outweighed risk-to-force concerns. In the long-standing customer-service model of space operations, risk has often been calculated in terms of risk to equipment (very expensive satellites), in terms of unintentional escalation, or in terms of intelligence gain/loss (Will this activity reveal something we do not wish to reveal?). Risk to force, however, has remained relatively low because most space operations forces, even forward-positioned ones, collocate with established military bases and thereby gain a measure of security. In a more expeditionary model, such forces could collocate with mobile fleets or field armies, raising the risk potential considerably but also promising greater operational flexibility and depth. The problem of determining risk for global space operations requires a model of its own, one that we will explore in Chapter 5.

Section 5: Combined Arms Dilemmas

With a few available frameworks in hand—the levels of war, principles of war, elements of operational art and design, and tactical symbology—the operational artist may now begin combining tactics with purpose. The synchronization of tactical action to achieve strategic objectives is the essence of operational art, and operational art will be essential in the battle beyond.

It is possible, of course, to combine tactical action to achieve tactical or operational objectives. In fact, tacticians do this implicitly. The armored brigade's maneuver does not succeed without artillery, aviation, and logistical actions. The primary purpose of these multiple arms is to present multiple dilemmas to the enemy such that the enemy cannot counter them all. Their secondary purpose is to retain optionality in the face of the enemy's available dilemmas. Certainly, the combination of multiple arms and services achieves a synergy beyond the sum of their individual efforts. The full incorporation of space warfare into the larger all-domain operations depends upon the notion of combined arms, a notion that requires the explicit understanding of the available arms, their functions, and the ways in which they combine to contribute to both the dilemma and the option sets.

Achieving combined arms activity requires more than simply fighting together with a variety of capabilities. It requires presenting to the enemy dilemmas and problem sets that the enemy must address, each as unpleasant as the next. In his book *Maneuver Warfare*, military theorist Bill Lind provides the quintessential example of a military dilemma. Combat engineers emplace a minefield in front of a supported infantry unit's battle position. The infantry unit emplaces direct-fire missile systems to overwatch the minefield. The combination of the mines and the missiles presents a dilemma to the enemy: Do we maneuver rapidly to avoid the missiles, or do we maneuver slowly to avoid the mines? Whichever choice the enemy makes will incur a cost, and neither choice is a good choice. A thinking enemy will seek to develop options that will increase his chances of success: modifying their avenue of approach, requesting additional assets like attack aviation or long-range fires, or choosing to attack under light and weather conditions that favor the attacker—just to name a few possibilities. Functionally, similar arms exist in the community of space warfare—satellite communications, missile warning, electronic warfare, orbital warfare, and others, each with their own subspecialties. In space warfare, similar potential exists for the creation of dilemma sets through the employment of the available arms. The viability of these dilemma sets depends as much upon the willingness of the institution to organize and equip for them as it does on the ability of the warfighter to train for and implement them. In addition, the employment of multiple phenomenology space weapon systems (lasers, kinetic, RPO methods, cyber, and so on) against a single target ensures the adversary becomes under attack from multiple avenues of approach. His satellite defenses against one space weapon phenomenology (e.g., lasers) might be irrelevant against a different kind of attack (e.g., kinetic kill). It should also be noted that many forms of satellite defense involve the targeted satellite going off-mission while defending or fleeing, and thus the attacker at least has accomplished the effect of temporarily negating the mission usefulness of the target even if he does not achieve the effects directly.

By the preceding definition, simply forcing an enemy to make a choice does not constitute a dilemma. The enemy may go right or left, may choose to attack or not to attack with various means, and friendly forces may do the same. Dealing with such choices may prove inconvenient, and it may alter established plans and timelines, but these things are to be expected in warfare. The essence of a dilemma is that it induces a state of psychological shock such that the enemy does not know which bad option to choose, losing any chance of achieving or maintaining the initiative. In other words, all dilemmas are composed of option sets, but not all option sets constitute dilemmas. In fact, while options sets are commonplace, true combined arms dilemmas are rare. In space operations, the missile warning/space domain awareness dilemma, the electronic warfare/satellite communications dilemma set, and the military/commercial dilemma set provide three practical examples to illustrate the principle.

The Missile Warning/Space Domain Awareness Dilemma

A temptation in the consideration of dilemma sets is to try building dilemmas out of like capabilities. One such example is in the complementary functionality of missile warning satellites and ground-based missile warning radars. While both contribute to the space mission area, for missile warning purposes they do not collaborate to form a dilemma set because they do not force a hard choice upon the enemy's option set. They do add robustness to the missile warning mission, which is valuable, but the dilemma comes not from a multiple-phenomenology missile warning architecture but from the fact that some ground-based missile warning radars can perform a secondary mission—that of space domain awareness.*

Because they emit radar energy, it is possible to use ground-based radars to track space objects, assuming that their energy outputs are sufficient to provide returns at orbital distances and assuming that their data processing allows for interpretation of such returns. In theory then, with enough contributed energy and with the proper processing suite, any radar could contribute to the space domain awareness mission. In a world where every friendly force radar in the inventory is contributing, a nation could effectively crowd-source both missile warning and space domain awareness through an interconnected network of sensors (network vulnerabilities notwithstanding). In so using radars for multiple purposes, a dilemma for the enemy emerges. The dilemma is this:

Do we attempt to interfere with an enemy's space domain
awareness network in order to remove their knowledge of friendly and enemy orbital assets?
Or
Do we allow an enemy to maintain their space domain awareness
because interfering with their SDA network requires interfering with
their missile warning network, which could risk rapid escalation?
Or
Do we risk conflict escalation and allied realignments if we attack out-
of-theater space surveillance assets that are commercial and located in
neutral countries but are contributing to space domain awareness?

This dilemma set takes advantage of a certain worldview, which if not already shared across all the major military actors of the world probably should be. That view is the one in which interfering with an enemy's missile warning sensors (or possibly even air warning radars) is a logical prelude to a

---

*   Duane Lankford, "$866M Contract Sustains Six Missile Warning Radars," Wight-Patterson Air Force Base, June 26, 2018, https://www.wpafb.af.mil/News/ArticleDisplay/Article/1559778/866m-con-tract-sustains-6-missile-warning-radars/.

nuclear first strike. Herein we see a dilemma set in which a tactical action—a cyberattack against a ground-based radar site, for example—could lead to definite strategic consequences. Indeed, space domain operations often lead to such near-immediate escalations, but it is important to remember the distinction between tactical and strategic and the fact that the operational level always exists, even if not obvious.

This dilemma set is only one part of the endeavor, however. The second part is exploring the implications of the dilemma set. The first implication is one that may seem obvious: deliberately linking missile warning functionality with space domain awareness functionality to the greatest extent possible, especially into radar systems that have not traditionally been used for this purpose. Such functionality provides redundancy to the network of missile warning radars and may cause an adversary to second-guess their particular targeting set. The second implication is that separate functionalities may not necessarily require the command of separate organizations. Through data lines of communication, multiple organizations may have access to the same data, parsing out the relative parts for their missions and leaving the rest on the digital scrap heap. This view is perhaps aspirational; sensor tasking authority is very real in today's military, but the future of big data promises such potential. The third implication involves the interplay of neutral actors who may be hosting multipurpose, commercially owned military infrastructure. Not only are the military considerations of space and missile defense sensors involved, but economic and diplomatic considerations come into play, complicating the decision-making process. A final implication is one that seems to translate across all space-domain dilemma sets: the need to deliberately manage data lines of communication, either through securing them, expanding them into zones, or reducing dependency on them (more on this in the final chapter).

The Electronic Warfare/Satellite Communications Dilemma Set

In the twenty-first century, electronic warfare has become a fundamental skill set for military operations in all domains. The doctrinal subsets of electronic attack, electronic protection, and electronic support (actions "to search for, intercept, identify, and locate or localize sources of intentional and unintentional radiated EM energy," which may support either attack or protection options) apply to the space domain as surely as they apply to the air or maritime domains.* In combining these functions, one may imagine a dilemma set that involves electronic attack (think direct-fire missile) and electronic protection (think minefield) capabilities.

Whether army, navy, or space force, many different kinds of units regularly monitor their battlespace to detect electromagnetic interference, whether unintentional or intentional. Even the military intelligence company organic to a brigade combat team has this capability. For space operations, the case is no different. Tactically, it should be possible to employ an electronic attack asset (i.e., a jammer) in coordination with an electronic support asset. The pair would function like an artillery battery and a forward observer team, one providing feedback to the other on the accuracy of its fires. If an electronic attack asset operates an organic capability to assess its own fires—and in so doing becomes a direct-fire asset—all the better.

Just as in the pair of missile warning radars and missile warning satellites, electronic attack and electronic protect assets are likely to work better in pairs. Also like the missile warning case, the attack/protect pair does not exactly contain a dilemma in itself. True, the enemy must choose whether to withhold capability in relative safety or to employ it and risk counter-attack, but this set of choices probably does not force a choice between equally unpleasant alternatives. That requires an external element, one that is conceptually different than electronic warfare: satellite communications.

Satellite communications leverage the electromagnetic spectrum as surely as electronic warfare does, and for that reason, the concepts of data lines and data zones of communication apply to the discipline of satellite communications as surely as they apply to the pairing of satellite communications and electronic warfare. The dilemma is this:

---

*   US Joint Staff, Joint Publication 3-51, *Joint Doctrine for Electronic Warfare* (Washington DC, 2000), 1–2.

Do I sacrifice effective command and control
of my subordinate elements?
OR
Do I expose my command-and-control elements
to the risk of geolocation and destruction?

To leverage satellite communications, the enemy must emit an electromagnetic signature, transmitting their signal to a satellite for relay to either another satellite or, more likely, another ground unit. In so doing, they open themselves to intercept or detection by friendly electronic support assets. These electronic support assets, then, may cue electronic attack (or other attack) assets to deny the enemy's ability to communicate via satellite. In a hypothetical space-only scenario, the electronic support (sometimes called a "defensive space control" asset) detects the jammer and cues an electronic attack unit to deny the signal that the enemy ground-control unit is using. In a multidomain scenario, an electronic support system detects and geolocates the enemy satellite jammer and cues a nearby grounds unit to destroy it.

As in the missile warning/SDA dilemma, the EW/SATCOM dilemma carries implications along with it. The first implication is that having electronic attack and electronic support functionality contained in the same system is a great benefit. The second is that, if attack and support functionalities exist in different units, their paired employment is a necessity during wartime. In peacetime, electronic support may operate independently for the purposes of monitoring and deconflicting the electromagnetic environment, but its benefit to combat operations exists only if paired with a shooter—electronic or otherwise. Third, SATCOM, more than any other space mission area, requires the deliberate management of zones; the enemy's anticipated electronic support and electronic attack operations will have a large say in how those zones are managed in theory and in combat. The corollary to this statement is that the enemy will have a plan for managing his lines and zones based on friendly support and attack orders of battle.

The fielded force, therefore, must be able to impose sufficient cost to the enemy's zones to be relevant to the war. Token forces are not only irrelevant, but they are actually a liability because they detract resources from more viable elements. A final implication—at the risk of overemphasizing a theme—is that satellite communications carry inherent vulnerabilities. As with any data line or zone, these vulnerabilities can be mitigated through securing them, expanding lines into zones, or reducing dependency on them.

The Military/Commercial Dilemma Set

In terms of numbers of satellites, numbers of ground stations, and consumption of the electromagnetic spectrum, satellite communications are the most ubiquitous form of space operation for both military and commercial applications. Commercially provided intelligence, however, is a growing field. Whether imagery, signals intelligence, or any other product, the potential for expansion in this area is significant. Regardless of the type of commercial space mission, however, the following dilemma set applies.

It is possible to employ, for military purposes, a combination of military capabilities and commercial capabilities to contribute to the conflict's objectives. In fact, for shipping and air transport, the US military does this all the time because it is impractical to maintain a fleet of sufficient size purely for military use. For satellites, the menu of capabilities is already significant and in the process of expanding. The dilemma is this:

> Do I allow the enemy military access to commercial capa-
> bilities, enhancing his warfighting capacity?
>
> Or
>
> Do I deny the enemy access to commercial capabili-
> ties and risk the economic and political consequences?

In this dilemma we see an example of a dilemma set that straddles the military strategic and the grand strategic. On the one hand, we may allow an enemy access to satellite capabilities, allowing their fighting force to be more effective. On the other hand, we choose to attack commercial assets, which even if used for military purposes, could create political backlash. In a historical example, the Anglo-American alliance created this dilemma for the German navy, shipping warfighting material in commercial ships like the *Lusitania*.

Section 6: Stratagems

stratagem
a: an artifice or trick in war for deceiving and outwitting the enemy
b: a cleverly contrived trick or scheme for gaining an end
The Merriam-Webster Dictionary

A stratagem, like a strategy, incorporates available means to achieve a desired end. A stratagem, however, is conceptually distinct from a strategy for two reasons. First, strategies are larger and more comprehensive in their scope, but stratagems usually involve only a limited number of means. Second, the hallmark of the stratagem is that the ways incorporate cunning, deception, or deliberately unexpected behavior. Stratagems, like dilemma sets, may serve as ready-made concepts for employment by the operational artist and the strategist, likely as part of a larger approach.

As in the case of combined arms dilemmas, it is not possible to list every form of stratagem that may be applicable to space warfare. The important thing, however, is to introduce the concept of stratagems to the students of space warfare, and through study of examples, they may begin formulating their own, adapting old ideas and creating new ones as the ever-dynamic realm of warfare changes. To that end, the rest of this section offers example stratagems.

Stratagem #1: Herd Space Personnel

Action: Destroy a significant portion of the enemy's space-related ground targets, leaving one ground segment facility operational. Assume that key space support personnel will converge to this lightly damaged site to conduct repairs or continue operations. At some point in the near future—perhaps twelve hours later—destroy the site and any personnel located there.

Effect: Destroy the enemy's most important space asset: key technically trained space operations personnel. This stratagem also sends a message to the international community that personnel supporting the enemy's space efforts may be at risk.

Stratagem #2: Herd Communications

Action: Selectively destroy or temporarily disrupt specific enemy space systems communications assets so that critical enemy sensor and command and control data is directed to known paths that friendly forces can monitor.

Effect: Make the enemy force vulnerable to intelligence exploitation. In his response, the enemy may reveal much about the way he responds when threatened. Their response should drive planning for future offensive action.

Stratagem #3: Funnel Communications

Action: Selectively disrupt enemy space system communications, encouraging the enemy to direct critical sensor and command and control data into paths with low data rates.

Effect: Delay the receipt of critical data among enemy headquarters while conserving space electronic warfare assets. This stratagem aims to frustrate the enemy's efforts and degrade his operational tempo—in Clausewitzian terms, to increase the fog and friction of war.

Stratagem #4: Herd Sensors

Action: Selectively deny enemy space sensors in order to temporarily remove their full space domain awareness capability. Allow the enemy to observe again when convenient for the friendly force, to either show them true or false dispositions.

Effect: Control the enemy's perception of the battle beyond. The friendly force may wish the enemy force to know its true disposition, if the advantage is overwhelming and the friendly force commander desires the enemy to know he is beaten. Conversely, the friendly force may present a false disposition if it is in a disadvantageous position or if space forces are party to a feint.

Stratagem #5: Hidden Disrupt

Action: Employ weapons with low probability of detection and attribution to minimize the chance of a negative international response. Temporarily disrupt adversary spacecraft operations at random times.

Effect: The adversary loses confidence in his space systems. Repeated disruption keeps him continuously off balance, especially if deliberately synchronized within enemy decision cycles.

Stratagem #6: Hidden Negate

Action: Employ weapons with low probability of detection and attribution to minimize the enemy's perception that counterspace operations have begun against them. Slowly increase the tempo of disruptions, starting with minor anomalies that are easily detectable, as a way to message the escalation of hostilities.

Effect: Similar to the Hidden Disrupt stratagem, the Hidden Negate stratagem aims to affect the enemy's confidence in his space capabilities. The Hidden Negate stratagem, however, lends itself more to open conflict. If the enemy has not employed its space systems in conflict before, their decreasing reliability under combat stress may initially be understandable and acceptable. As the conflict escalates, however, this negation becomes increasingly costly, especially if the enemy is highly reliant on space capabilities.

Stratagem #7: Periodic Degrade

Action: Synchronize the use of degrade-type weapons to correspond with the reconstitution or replacement time of the target's capability.

Effect: As the enemy starts bringing alternate space capabilities online (perhaps on-orbit spares, perhaps responsively launched replacements), the friendly force degrades them at once, inhibiting their ability to effectively reconstitute. This minimizes enemy space systems employment but has a relatively low value in terms of operational shock.

Stratagem #8: Rolling Disrupt

Action: Temporarily disrupt space assets of the enemy's supporters for short lengths of time, periodically moving on to a different asset. Use low probably of detection/attribution weapons to give the impression of reliability issues, not intentional attack.

Effect: Supporters of the enemy may not necessarily be your enemies. By this stratagem, your enemy's supporters may be kept guessing as to the ultimate fate of their space systems if they continue to offer their support to your enemy.

Stratagem #9: Sweep the Ground

Action: Destroy all enemy space-related ground targets with a minimum of collateral damage. As in the tactical task definition, destroy here means to render combat ineffective until reconstituted.

Effect: With all ground sites destroyed, the enemy cannot task or download data from its satellites. The cunning in this stratagem stems from the consideration of the postconflict environment. A belligerent may more easily replace ground equipment than on-orbit equipment, so rendering ground stations ineffective may be more politically acceptable than rendering satellites ineffective. Acceptability is more likely if the destruction is difficult to attribute as through a cyberattack on ground infrastructure.

Stratagem #10: Sweep the Skies

Action: Destroy all enemy satellites, whether military, civil, or commercial, nearly simultaneously so that enemy protective and reconstitution measures cannot be implemented effectively.

Effect: Although likely to induce operational shock on a space-dependent enemy, this stratagem is only viable if hostilities are imminent. Depending upon the size and scope of the enemy's fleet, the approach is likely to be very resource intensive and will require near-global synchronization. For its difficulty and severity, such a course is also the most unexpected. However, because of the vagaries of orbital dynamics, there are only two types of massed attacks on orbit. The first is to maneuver the attacking satellites simultaneously. However, these ASATs will reach their targets at different times. The second technique is to maneuver the attacking satellites at differing times so they all reach their targets simultaneously for maximum shock effect. However, both of these techniques provide the defending country with the potential for strategic warning, especially with good SDA systems. Massed attacks against the electromagnetic spectrum and ground segments are also possible.

Stratagem #11: Resource Reduction

Like strategies, stratagems may be direct or indirect; the indirect approaches requiring a much greater time investment. In this stratagem, an adversary attacks space systems that are not that important to the current conflict. This draws out enemy intents and wears away at his available resources (including ASATs, maneuvering fuel, space surveillance and intelligence capacity, tracking assets and command personnel and communications channels). This scenario may exist unintentionally in the real world because of the difficulty of assessing the value of space assets. An adversary will likely rank the value of his enemy's space assets differently than the friendly force ranks them, and vice versa. For this reason, an adversary probably will attack some space assets that will leave you scratching your head for a reason why. After all, friendly intelligence collection capabilities are imperfect. Since a major space war has not occurred yet, it is difficult to foresee which space assets will prove more valuable than others. The value of systems will likely vary by theater, but prudent planning will provide insight into this question, guiding the implementation of this stratagem, even in peacetime. In peacetime, a resource reduction stratagem may be a decades-long commitment and would most benefit a belligerent who is able to maneuver their satellites with the greatest efficiency (or perhaps refuel and repair them in space). Or one might just covertly attack space systems supply chains.

Conclusion

The operational level of war ties the strategic level to the tactical level. As the schema at the beginning of the chapter suggests, the activities involved in the operational level focus on understanding what is happening and developing approaches and courses of action that synchronize the tactical actions of many subordinate elements to achieve strategic objectives. Visualizing such approaches—both in the minds of the participants and through command-and-control systems—requires common terms and the ability to depict what is happening graphically. The sketches and screenshots presented in this chapter offer a handful of visualization options, some being more useful for conceptual planning, others for monitoring and making sense of current operations. All have the potential of contributing to the synchronization of tactical action.

If there is a visual vocabulary to the operational level—one that takes advantage of the tactical vocabulary developed in the previous chapter—there is also a verbal vocabulary. The elements of operational art and design allow operational-level practitioners to focus on common terms of art. In fact, objectives, decisive points, lines of operation, and the other elements are means that may be used to create ways.

Like the principles of war, the elements of operational art and design provide useful concepts for practical application in military planning of all kinds. Their utility for space operations, therefore, should not come as a surprise, but in exploring the doctrinal definitions of each, it became apparent that, the concepts themselves are not rigidly set in stone. Meanings of terms change over time, connotations of verbiage change, and practitioners of different stripes come to favor slightly different meanings. As useful as these concepts are, then, they cannot strictly be said to remain immutable, but this characteristic is not a shortcoming of doctrine. On the contrary, a synoptic approach allows people to ponder the nuances and ask of themselves, "What exactly do we mean by this?"

Dilemmas and stratagems also provide useful tools for the operational artist. Unlike elements, however, dilemmas and stratagems exist in ready-made combinations of means and ways. Clearly not every combination of tactical activities provides the opportunity to create an intractable dilemma for the enemy, but this realization should discourage us neither from seeking dilemma sets nor from employing other creative tactical options. It is, as a matter of tactical opinion, good practice to employ electronic attack units in coordination with electronic support units. If this activity contributes to the attainment of a strategic objective—wholly or in part—we may consider the combination of tactics a matter of operational art. If the combination does not contribute to the attainment of a strategic objective, this is no great matter. The activity must certainly be useful (otherwise, why would we be doing it?), but we would more properly label it as tactics—more expansive and complicated tactics, but tactics nonetheless.

It is here worth addressing again whether, in space operations, tactical actions lead directly to the attainment of strategic objectives without an intermediary operational level. Chapter 1 noted the tendency to confuse strategic actions with tactical actions of strategic consequence. An intelligence satellite may change orbit, telegraphing a strategic message to an adversary, but the act of relocating the satellite itself is a tactical one. The fact that the satellite is a national-level asset collecting on strategic-level objectives does not change this fact. Similarly, while the tactical action is the visible one, this singular act did not become accomplished without the coordinated tactical efforts of multiple units and organizations, many of whom are not readily associated with one another.

The point of this diversion is that tactical actions that contribute to strategic objectives do not exist independently from other tactical actions. They require the support of multiple other tactical activities, which, in a data-enabled operating environment, may be transparent to the unit that is the main effort. In this example, the question we must ask is: what units are working to make the move of an intelligence satellite possible? First, there are the people responsible for space domain awareness. There are likely multiple units involved in this effort, operating a variety of sensors from radars to ground-based optical sites that will determine if movement of this satellite will cause a collision with other satellites—potentially a threatening act—or collide with space debris and destroy a very expensive satellite. This effort feeds a fusion cell that is making sense of the orbital picture—possibly an intelligence unit. This intelligence unit may also be studying the enemy's activities to assess their likely courses of action and determine how the enemy would respond to the relocation of the intelligence satellite in question (such as covering up sensitive activities on the ground that this satellite can observe). This assessment makes its way to the organization responsible for the operation of the satellite, which determines, based on their higher guidance, to relocate it. The sensor units receive a tasking to verify the relocation of the intelligence satellite and to determine if any suspicious activity is taking place on orbit. Meanwhile, the intelligence unit begins its assessment of the enemy's response.

It is temptingly convenient to view a B1 bomber nuclear strike or a special operations raid against a high-value target as merging the tactical into the strategic, but such simplicity is, from a warfighter's perspective, disingenuous. In both cases, the bomber and the raiders are likely to depend upon intelligence collection and analysis, command and control activities, logistics support, deception operations, and cyber activity—to name a few—to get to the point in the operation where their tactical operation has the possibility of succeeding. The coordination of these activities to achieve that strategically important moment is operational art.

Such understanding is particularly important for practitioners of space warfare who, in large part, are still trying to create a common language. In the past, this language creation has often focused on the creation of space-unique terms, engendering a lexicon that stood apart from more established doctrine. As discussed in other parts of this work, such uniqueness may sometimes be necessary, but only after other established concepts and metaphors have been thoroughly explored. This is the process by which institutions grow in thought and ultimately become effective. For Sir Julian Corbett, theory was just this, the "process by which we coordinate our ideas, define the meaning of the words we use, grasp the difference between essential and unessential factors, and fix and expose the fundamental data on which everyone is agreed."* It is possible that not everyone will agree, and it is possible that the existing elements will require expansion or contraction for space-specific scenarios. Since the acme of operational art, however, is to combine the means of all domains to meet strategic ends, the elements must remain a common part of the joint force's shared vocabulary. What's more, the thought must be accomplished before the next conflict so that we may trade brain cells for blood cells at a favorable exchange rate.

Vignette: A Troubled Peace

It is nominally peacetime. The nation of Newmex has been experiencing growing tensions with its neighbor Califon over regional trading disputes, but neither side has openly declared hostilities. Military activities on orbit and within the electromagnetic spectrum are constrained by the loosely defined norms, which Newmex and Califon largely share. Both nations have comparable military capabilities across all domains, and although neither enjoys state-of-the-art space capabilities, they each have a wide array of means available by which to pursue national objectives on orbit, across the electromagnetic spectrum, and on the earth's surface. As part of their efforts to deter aggressive Califonian orbital activity, the grand strategy of Newmex has recently dedicated significant assets to improving its space domain awareness network through the addition of ground-based radars and optical sensors positioned both within Newmex and within the territories of strategically important partners.

The emplacement of these sensors required deliberate analysis. The following SWAT screenshot shows a handful of Newmex's space sensor

---

*   Julian Corbett, *Some Principles of Maritime Strategy*, in *Roots of Strategy (Book 4)*, ed. David Jablonsky (Mechanicsburg, PA: Stackpole Books, 1999), 164.

requirements as they nest with operational priorities. Newmex ranked their military requirements and optimized them to determine which sensors were needed, what their revisit times and sensitivities would need to be, and where they should be located. Figure 81 and Figure 82 depict this analysis, the first showing the logical flow of the analysis as part of a series of lists and drop-down menus, the second showing the outputs of the analysis as a series of graphs. Figure 83 provides a blow-up of the central bar graph of the preceding graph and provides a recommendation on which faces of a satellite the new sensors will need to observe.

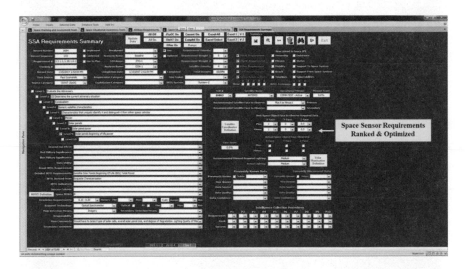

Figure 81: SDA Optimization Schema. *Source:* Graphic created by Paul Szymanski using the Space Warfare Analysis Tool (SWAT) that he developed.

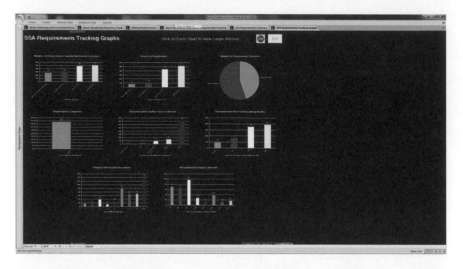

Figure 82: SDA Optimization Schema—Statistical Results. *Source:* Graphic created by
Paul Szymanski using the Space Warfare Analysis Tool (SWAT) that he developed.

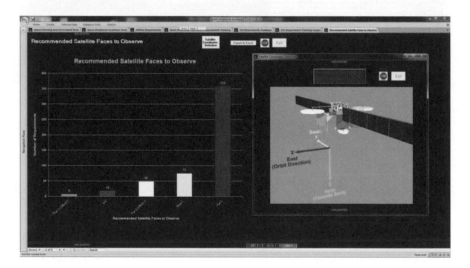

Figure 83: Recommended Satellite Faces to Observe. *Source:* Graphic created by Paul
Szymanski using the Space Warfare Analysis Tool (SWAT) that he developed.

While the SDA infrastructure initiative is of extreme importance to the
majority of the Newmex government, even the best acquisition and procurement efforts of the Ministry of Defense cannot deliver capabilities at
the speed of need. Indeed, the prolonged period of peace in Newmex has

tipped the balance between the institutional and the warfighting aspects of defense; rather than having warfighting needs drive the institution, institutional processes have come to dictate what the warfighters have available. Fortunately, however, the Newmex military has invested significant efforts into its development of nonmateriel solutions. The need for viable contingency plans that incorporate robust space responses has inspired critical and creative thought across the military. Although many of the ideas remain as "best practices," new concepts are emerging with every wargame, and the force is making deliberate efforts to include these lessons learned in its education, training, and doctrine development programs. In addition to the ideas already discussed in this book, Newmex has made significant progress on how it thinks about weapons release authorization levels, intelligence indicators, and rules of engagement as they pertain to space warfare. Even this work, however, may not be sufficient to overcome the persistent and growing threat of Califon's space capabilities.

Weapons Release Authorization Levels

Since the operating environment is peaceful, the Newmex military is enforcing positive control of space weapons against targets that are supporting enemy activity in the area of responsibility—per their standard weapons release authorization rubric, which they developed based on historical air defense doctrine of the United States (Figure 84). As such, positive identification, tracking, and a valid situation assessment of spacecraft activity within a designated space defense area are their primary operational activities. If the situation escalates either in space, on earth, or in the combined operating areas (terrestrial and space), the Newmex military is prepared to adjust weapons release authorities for increased operational responsiveness. In fact, in accordance with anticipated enemy courses of action, Newmex has already weaponeered their initial response and could be ready to execute on orbit, within the electromagnetic spectrum, and from the ground within seventy-two hours of approval.

Level of War	Weapons Release Authorization Level				
	Space Positive Control	Space Autonomous Operation	Space Weapons Hold	Space Weapons Tight	Space Weapons Free
Peace	Yes	No	No	No	No
Space Crisis	Yes	Maybe	Maybe	No	No
Conventional Terrestrial	Yes	Yes	Yes	Maybe	No
Conventional Terrestrial & Space	Yes	Yes	Yes	Yes	Maybe

Figure 84: Standard Weapons Release Authorization Rubric. *Source:* Created by Paul Szymanski.

Apart from positive control, there are four other categories of weapons release authorizations, three derived from common military usage (hold, tight, and free) and a fourth category (autonomous operations) to account for the possibility of independent action on the part of the satellite without human intervention. Briefly, in weapons hold, weapons systems may only engage in self-defense or in response to a formal order. In weapons tight, weapons may be fired only at targets positively identified as hostile. And in weapons free, operators may engage any target *not* positively identified as friendly. The category of autonomous operations may technically encompass hold, tight, or free protocols. When operating autonomously, a space system has lost communication with human controllers. Based on its on-board surveillance capabilities and its onboard logic (i.e., its artificial intelligence), it may engage hostile targets independently of human action. This automatic state may occur on a regular basis because of orbital movements outside regions of ground coverage and control. For the time being, however, positive control is the appropriate posture given the state of conflict (peace) within the AOR.

If the situation begins to escalate in a different AOR, Newmex has developed a contingency for that too. In fact, the Newmex military has developed contingencies for responses in each of the AORs where their national interests could be threatened. In these cases, weapons release authorizations are even more restrictive, requiring change authorizations at higher levels of command. These restrictions acknowledge the potential strategic risk in launching an attack from one relatively peaceful theater into another that is in a state of conflict. Of course, as with all military rubrics, commanders may need to modify responses as situations develop, keeping their higher headquarters informed of their intentions and avoiding any potential oversteps of their granted authorities. Figure 85 lays out their envisioned weapons release authorities for such a scenario. As in the case of an inter-AOR conflict, a shared understanding exists among the terrestrial and space combatants for what authorizations exist at each stage of conflict.

Level of War	Weapons Release Authorization Level				
	Space Positive Control	Space Autonomous Operation	Space Weapons Hold	Space Weapons Tight	Space Weapons Free
Peace	Yes	No	No	No	No
Space Crisis	Yes	Maybe	Maybe	No	No
Conventional Terrestrial	Yes	Maybe	Maybe	No	No
Conventional Terrestrial & Space	Yes	Yes	Yes	Maybe	No

Figure 85: Weapons Release Authorities for Conflict Spanning AORs. *Source:* Paul Szymanski, Outer Space Warfare: Restraining Conflict," *Defense & Security Alert*, vol. 12, issue 01 (October 2021): 37. https://www.dsalert.org/DSA-Editions/Oct_2021_Paul_S_Szymanski.pdf.

Intelligence Indicators

Despite the relative peace, Space Defense Region "GEO Asia," a subdivision of Space Defense Area "GEO" (SDA-GEO) remains full of potentially hostile activity. While it would be desirable to provide thorough assessments of all space defense regions, Newmex simply cannot afford the significant investment in intelligence capacity that such a course would require. In fact, with the support for the space domain awareness expansion secured across the whole of its government, the Newmex military has recently realized that, under its current force structure, it will not have enough trained intelligence personnel to support the new hardware. Software algorithms may reduce some of that burden, but software acquisition presents its own difficulties. A more thorough analysis of capability gaps might have allowed Newmex to avoid this shortcoming, but with the limited current intelligence capacity available, Newmex's space operations forces are doing their best, focusing their efforts on the biggest potential threats of SDA-GEO.

The current activity assessment within the space defense area (and at a higher echelon within the entire space defense region) is informed by a number of semistandardized threat warning and assessment indicators, which help inform decision authorities about possible enemy courses of action. Again, based on intelligence capacity, it is not possible for Newmex personnel to assess all indicators (see Figure 71 for an example of an automated intelligence assessments for space), but they have been well trained on the envisioned operational scenarios and have high confidence that they are familiar with Califon's systems and tactics. Similarly, strategic-level intelligence organizations are closely monitoring the activity of Califon's space industry, particularly activity at a handful of research and development facilities where in recent months, parking lots have been fuller than normal. Although it has been difficult to confirm, the increased power usage, local housing demand, and observed equipment tests (along with a number of other indicators) suggest that Califon has invested significantly in their development of a mobile, direct-ascent satellite missile.

Despite the importance of the strategic intelligence, the operational and tactical forces are focusing on capabilities closer to their space defense regions and sectors. A possible enemy antisatellite weapon, the Savonne micro inspector satellite is operating unusually close to a Newmex communications satellite called the Itatingui Large. The Savonne belongs to Califon, and although some information is available about the Savonne and its place in Califon's war plans, the space domain awareness coverage of it is episodic, often with significant time gaps between sightings. To make matters worse, the microsatellite is small enough to be barely discernable by Newmex's ground-based radars, so space operators cannot be sure whether the satellite has moved or, given a change in orientation, the sensors have now reached the limit of their effectiveness. Despite the efforts of multiple domestic and allied intelligence agencies, there is still a fair amount of doubt among Newmex forces about the exact capabilities of the Savonne.

Regardless of local Savonne activity, the Itatingui has continued to operate nominally—that is, until recently. During a period of intense solar activity, the Itatingui experienced several outages caused by unusual upset events and dysfunctional behaviors within some of its internal systems. The result was a significant degradation in the power output of the satellite's main antenna, a reduction of about 20 percent. During these troubles, the Itatingui satellite was supporting Newmex military exercises in the western Pacific. The Savonne inspector satellite may have been nearby, but operational forces have been unable to verify through organic space domain awareness assets, and it remains uncertain whether the upset conditions experienced by Itatingui were the result of natural phenomenon (i.e., the solar storm) of or the approaching enemy satellite. Because of this uncertainty and the currently low conflict level (peace), Newmex's space command opts only for passive protective options that will not significantly impact their satellite's mission. Importantly, this decision was not for a lack of critical analysis. The command had, through a series of extensive wargames and simulations, developed a thorough menu of possible reactions to a variety of enemy courses of action (Figure 86). In the end, the commander made the determination based on the best information available.

Time Sequence	Category	Actor	Target	Escalation Ladder	Probability of Occurrence	WBS	Action Reaction
35	Satellites	Califon	Newmex	P.A.A.1	8	N.S.R.3	Newmex Bicudo Large LEO Photo Satellite is permanently partially blinded when over flying the disputed oil fields
35A	Political	Newmex	Califon	P.1.C.0	8	N.S.R.3.0	Do nothing to increase escalation ladder
35B	INTEL	Newmex	Califon	P.1.A.0	10	N.S.R.3.1	Determine if degradation is caused by natural events, equipment failure or human actions, whether intentional or unintentional
35C	Forces	Newmex	Califon	P.1.C.0	9	N.S.R.3.2	Increase military alert level (DEFCON)
35D	Ground Stations	Newmex	Califon	P.1.A.0	9	N.S.R.3.3	Contact other Newmexian space-related ground facilities to determine if multiple ground outage incidents are occurring
35E	Satellites	Newmex	Califon	P.1.A.0	9	N.S.R.3.4	Contact other Newmexian TTC ground facilities to determine if multiple satellite outage incidents are occurring
35F	Satellites	Newmex	Califon	P.1.A.0	9	N.S.R.3.5	Check with Newmexian supreme military command to determine if other military incidents are occurring to Newmexian and allied forces
35I	Space Surveillance	Newmex	Califon	P.1.B.0	10	N.S.R.3.8	Increase surveillance and tracking for new and suspicious space objects
35J	Satellites	Newmex	Califon	P.1.B.0	10	N.S.R.3.9	Increase mission identification and country of origin determination for new and suspicious space objects (Space Object Identification - SOI)
35K	Satellites	Newmex	Califon	P.1.B.0	10	N.S.R.3.10	Increase signals intelligence collection on new and suspicious space objects
35L	Satellites	Orgonia	Califon	P.1.B.0	10	N.S.R.3.11	Maneuver Orgonian Abragh Nano LEO Inspector Satellite close to Newmex Bicudo Large LEO Photo Satellite for close inspection to help determine origin of mission degradations
35M	Satellites	Newmex	Califon	P.1.B.0	9	N.S.R.3.12	Increase satellite imagery, OPIR and RADAR surveillance and signals intelligence collection of Newmexian border areas
35N	Satellites	Newmex	Califon	P.1.B.0	8	N.S.R.3.13	Increase satellite imagery, OPIR and RADAR surveillance and signals intelligence collection of Newmexian internal areas
35O	Satellites	Newmex	Califon	P.1.B.0	9	N.S.R.3.14	Increase satellite imagery, OPIR and RADAR surveillance and signals intelligence collection of internal Califon activities
35P	Satellites	Newmex	Califon	P.1.B.0	9	N.S.R.3.15	Increase satellite imagery, OPIR and RADAR surveillance and signals intelligence collection of Califon **allied** activities
35Q	Forces	Newmex	Califon	P.1.A.0	9	N.S.R.3.16	Increase critical infrastructures defenses and surveillance
35AG	Political	Newmex	Califon	P.1.C.0	5	N.S.R.3.32	Cutoff diplomatic relations with Califon
35AP	Political	Newmex	Califon	P.1.B.0	9	N.S.R.3.41	Increase world attention to the problems of orbital space debris in order to slow down Califon's launching of new satellites
35BB	Political	Newmex	Califon	P.1.A.0	10	N.S.R.3.53	Engage in negotiations for space treaties and mutual defense pacts with other countries to increase space defense protection
35BC	Political	Califon	Newmex	P.1.A.0	10	N.S.R.3.54	Publically declare that any use of space weapons against Newmexian satellites will have a corresponding attack on the aggressor's space facilities associated with this attack, whether they be research centers, launch facilities, space surveillance sites, or command and control centers
35BD	Political	Califon	Newmex	P.1.B.0	9	N.S.R.3.55	Publically declare that any use of space weapons against Newmexian satellites will have a corresponding attack on the aggressor's **and their allies** space facilities associated with this attack, whether they be research centers, launch facilities, space surveillance sites, or command and control centers
35BE	Forces	Califon	Newmex	P.1.C.0	8	N.S.R.3.56	Initiate multiple false starts, threatening space and terrestrial maneuvers, etc. to induce your adversaries to begin constant satellite maneuvering, so as to waste their on-board fuel reserves before actual conflict starts
35BF	Forces	Califon	Newmex	P.1.C.0	8	N.S.R.3.57	Initiate random military orders, communications traffic, re-deployments and satellite maneuvers to confuse potential adversaries of your immediate plans and goals
35BG	Forces	Califon	Newmex	P.1.C.0	7	N.S.R.3.58	Launch or maneuver a new mysterious satellite that comes close to critical Califon satellites, to make Califon pause in its military execution plans, to show resolve, and as a warning to Califon to back down
35BH	ASAT	Califon	Newmex	P.1.B.0	10	N.S.R.3.59	Jam Califon propaganda broadcasts from their communications satellites directed at Newmexian dissidents
35BI	ASAT	Newmex	Califon	P.1.C.0	10	N.S.R.3.60	Initiate operational deployment of Newmexian Anti-Satellite (ASAT) systems

Figure 86: Possible Space COA Reactions. *Source:* Created by Paul Szymanski.

Rules of Engagement (ROE)

For the Newmex commander, one inspector satellite—to whose activities Newmex had become fairly accustomed—did not constitute a significant enough anomaly to raise the alarm. As such, the standing rules of engagement—another effort into which Newmex had put significant effort recently—only allowed for routine deception operations. Although he would be criticized for his decision later, the commander, assisted by his staff, did not see any other indicators of the expected Califon courses of action. There were, for example, no indications that Califon was modifying the altitudes or inclinations of any of its other satellites, no anomalies were experienced on any other friendly systems, and the staff confirmed that there were no known enemy naval or air operations at the time of the Itatingui anomaly—activities that would have potentially suggested massing in preparation for an operational-level or strategic-level offensive. However, seeing none of these, he assessed that the situation was still peaceful—not a space crisis, terrestrial crisis, or a combined terrestrial/space crisis. That being the case, the commander chose to not recommend an elevation of the conflict assessment and weapons release rules.

The choice to maintain the current conflict level label does not indicate inaction on the part of the commander, but it does indicate the choice to accept limited options. Figure 87 shows the active measures permitted under the standing ROE for peacetime.* Within this construct, direct defensive actions that deny, degrade, or destroy enemy space forces are forbidden; deception is permissible, and disruption attacks may be possible. Active operations may include the use of antisatellite weapons systems, electronic warfare, special operations forces against terrestrial targets, and other available weapons not primarily used in a space defense role. As one might expect, the ROE rubric goes hand in hand with the weapons authorization rubric. As the environment becomes more permissive for weapons release,

---

* The authors acknowledge that, in real life, rules of engagement are significantly more complicated than presented here, requiring much more legal nuance than we are capable of giving. We leave that to the lawyers.

the number of ways in which those operations may be conducted also expands—at least for the weapons systems available to Newmex.

Level of War	Rules Of Engagement (ROE)				
	Deception	Disruption	Denial	Degradation	Destruction
Peace	Yes	Maybe	No	No	No
Space Crisis	Yes	Yes	Yes	No	No
Conventional Terrestrial	Yes	Yes	Yes	No	No
Conventional Terrestrial & Space	Yes	Yes	Yes	Yes	Yes

Figure 87: Rules of Engagement for Various Levels of War. *Source:* Created by Paul Szymanski.

While active operations are limited, passive defensive operations are always permissible, and this was the area in which the Newmex commander chose to focus the efforts of his staff. Passive defensive measures are those measures taken to reduce the probability of and to minimize effects to space systems caused by hostile action. Many times, this includes attacks that are not detectible as human-caused and nonattributable to the attacking nation. Keeping in mind the need to apply them across ground, electromagnetic, and orbital segments, these measures may include camouflage, deception, dispersion, and the use of protective construction and design. Importantly, they may also include modified intelligence techniques, increased force protection levels, and modifications to unit training and exercise activities and information operations. Thus, choosing a passive defensive approach does not necessarily preclude seizing the initiative. Newmex, for example, may seize the initiative within the bounds of the intelligence and protection warfighting functions or within the information space. The ROE at this time, however, precludes it from seizing the initiative in the maneuver and fires warfighting functions. In the realm of the military strategic, Newmex space forces may initiate show-of-force activities, increase defense support to public diplomacy, or expand its presence in nonoperational areas.

Initial Assessment

It cannot be said that Newmex's assessment was overly tactical. Indeed, the Newmex commander assessed the tactical situation as he understood it, took into account the possibility of coordinated operational-level activities—even considering the possibility of coordinated tactical activity across multiple domains—and clearly had an appreciation for the military-strategic and the role of his forces within existing plans. In a word, it appears that the Newmex commander and his staff did everything right within the span of their considerable abilities. They took into consideration the available information and constructed a sensible narrative to reasonably explain what had happened and to build an appropriate response. Interestingly, however, Newmex formed an incorrect narrative.

Focusing predominantly on the orbital aspect of Califon's activities, they had not fully considered Califon's potential for activity in the electromagnetic spectrum. The inquiry at the end of the war assessed that this shortcoming was a direct result of the shortage of intelligence capacity within Newmex space forces. Although Califon had been operating the Savonne near to the Itatingui, they waited until the period of high solar activity to approach their target. The anomalies experienced onboard the Itatingui, however, were not the direct result of the Savonne's proximity operations. They were, in fact, the result of a ground-based electronic warfare asset that was being tested against the Itatingui. The Savonne was the reconnaissance and assessment asset, not the attack asset, and its tactical actions were synchronized with the electronic warfare asset. What's more, Califon sequenced its ground-based electromagnetic activity with peak solar activity to mask its operations.

While Newmex did have some capacity to monitor very small fractions of the electromagnetic spectrum for interference, what it had available was simply insufficient when compared to the robustness of its numerous data zones of communication. When Califon launched its degradation attack against the Itatingui, that data line of communication under attack was lost within the sea of other lines. Interestingly, as the postwar inquiry determined, Newmex's cyber forces had suspected that Califon was gearing up for an attack against a space target, but that information had never been shared with Newmex's space forces. Institutionally, they belonged to separate commands, which inhibited the warfighting integration between commands focused on different but highly interdependent domains.

In light of what they knew—or of what they believed they knew—Newmex leadership decided on a perceived space threat level. The commander decided that the status quo should be maintained, and although he increased intelligence, surveillance, and reconnaissance operations and expanded force protection activities, his response was conservative. An alternative option for the commander would have been to declare a space crisis (or to appeal to the appropriate authority to make such a declaration). Such a declaration, in accordance with established weapons release authorities and rules of engagement, would have automatically provided more military options should a response become necessary. It might also have increased the probability of space escalation as the adversary would probably detect this heightened military level through its intelligence assets.

In another approach, the commander could have left the "peace" assessment in place while simultaneously seeking precise modifications to the ROE to achieve specific ends. For example, even though it was peacetime, a destruction attack on the Savonne might have been a viable option if such action was sufficiently difficult to attribute. It might even have been possible to persuade an allied partner to engage in such activities. Such actions, particularly in peacetime, have potentially serious political implications and would likely require authorization from the highest levels of government. Furthermore, it would be wise to understand possible enemy counteractions to such an attack, including most likely and most dangerous courses of action, along with a consideration of the possible collateral damage.

Other nondestructive options were also available. Disturbing the Savonne satellite's attitude and orbital approach might have nullified the threat or even rendered it unusable for future action, though this latter course might also have led to subsequent collisions and potentially severe consequences. If feeling Machiavellian, Newmex could have even orchestrated a collision and publicly blamed Califon for its aggressive and irresponsible behavior in space. The political gain from such a tactic might be worth the loss of the Newmex satellite, particularly if it's at its end of life on orbit.

At the grand strategic level, more options were available in advance of the conflict, particularly with respect to agreements and commercial activity. Bilateral and multilateral agreements were very sparse, only offering the most modest access to shared imagery and signals intelligence from a few partners. What's more, established relationships between Newmex and any commercial partners were nonexistent, with its native commercial sector woefully underdeveloped. Califon held the advantage in these categories, receiving more varied and responsive imagery from its commercial partners—some of which were domestic—than Newmex was able to obtain from its allies. To add insult to injury, Newmex's diplomatic demarches for the Savonne's activities received no strong attention from the international community.

CHAPTER 5

Revisiting the Military Strategic

The operational vignette preceding this chapter provided a relatively simple scenario involving a small number of state actors and a handful of tactical actions sufficient to demonstrate operational art. For the fictional countries of Newmex and Califon, this scenario may have approximated the limits of complexity in their space operations. For a country like the United States with a continuous, global military presence, however, the set of options and dilemmas is significantly more expansive. Of course, no country or military has ever enjoyed limitless resources, and although US resources are vast, leaders and commanders at all levels of government must make allocation decisions. These decisions return us to the interface between the operational level of warfare and the grand strategic realm—the realm of the military strategic. Here, the most senior government officials, military commanders, and staff members assess risk to current military operations, advocate for and source future requirements, and integrate military-external resources across global, complex problem sets.

Making sense of these three activities and determining and developing an actionable strategy is no small task. One framework for thinking about such a challenge, however, is based on the previously discussed theoretical principle of the data line of communication. Through this concept, we might assess how Space Command (and potentially other commands) could use existing theater-level plans to pinpoint opportunities for assistance from other combatant commands, government agencies, nongovernment organizations, commercial entities, and allies, informing grand strategic options. By extension, the distinctly military capability gaps that emerge from this planning will inform the military-strategic efforts of the services in their capability development efforts and reduce risk for the joint force.

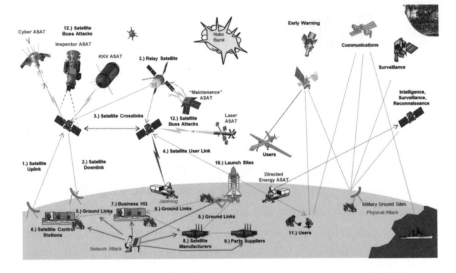

Figure 88: Space Warfare Possibilities. *Source:* Created by Paul Szymanski.

Figure 88 shows some of the possible means of space warfare and hints at the ways in which they are interconnected. In tracking the lines, it becomes clear that the space warfare quickly extends beyond the limits of any one commander's area of responsibility and even beyond the military's responsibility or capability to solve all potential problems. This is the reason that any military strategy necessarily requires the complementary efforts of a grand strategy—or at least the efforts of grand-strategic players—that leverages all available resources. The recent reestablishment of Space Command among other whole-of-government initiatives like the realignment of space traffic management authorities to the Department of Commerce highlights a growing emphasis on space operations and the need for more effectively integrating among grand-strategic partners. Within the military, space operations planning among multiple commands remains a challenge for the entirety of the joint force with Space Command serving as a focal point for space operations planning and expertise.

For Space Command, the planning challenges are daunting. As they are a new combatant command without an established family of plans, planning efforts must necessarily address three related facets of planning: how Space Command develops its own contingency plans, how the command fits into the existing plans of other combatant commands, how the command contributes to integrated plans that already involve multiple combatant commands, how it leverages the efforts of other government agencies, and how it will prioritize the capabilities that it provides both now and in the future—likely as a part of its campaign planning process.* To further complicate matters, Space Command has some authority to acquire materiel solutions outside of service channels and the lion's share of space expertise with which to inform other capability development efforts (like the creation of doctrine). These unique characteristics place Space Command into both the warfighting and the institutional camps of the military strategic realm while it is also functioning as a command responsible for overseeing operational-level activities.

As Space Command is a combatant command with global responsibilities, other combatant commands like European Command and Indo-Pacific Command would ideally include well-developed space operations activities into their existing plans. While existing plans certainly include space, the creation of a new combatant command and service for space have elevated the awareness of space and the need for novel approaches. Unfortunately, risk to space operations forces and capabilities cannot wait on the formalization of a new series of plans or the cyclical revision of existing ones through established institutional processes. In the interim, a bridging approach is necessary to provide structure to the planning efforts, to mitigate risk through resource prioritization, and to inform capability development.

---

*   There are, of course, multiple sources of strategic guidance for prioritization of effort, including the National Security Strategy, the National Defense Strategy, the Defense Planning Guidance, the Global Forces Management Implementation Guidance, and the Joint Strategic Capabilities Plan.

The crux of this approach exists at the operational level, which links the objectives outlined in combatant command planning documents to the analysis of tactical employment of space operations forces. Through a consideration of the operational level, the challenges of the military-strategic level become more tangible—a desirable outcome but one that is particularly desirable when dealing with a domain as intangible as space. A formal analysis with the concepts of the data line of communications (DLOCs) and operational-level decisive points at its center provides a useful way of linking the military-strategic to the tactical within the context of one plan; of prioritizing limited resources across multiple plans; of considering opportunities for support from whole-of-government partners, Allies, and commercial actors; of developing useful approaches for force development; and of ultimately reducing the risk to the joint force and its missions through the deliberate management of these data lines.

A Highly Simplified, Single OPLAN Scenario

Whether the supported or the supporting commander, combatant commands with global missions (Cyber Command, Space Command, Special Operations Command, Strategic Command, and Transportation Command) must prepare to execute multiple plans, possibly simultaneously. Furthermore, these commands must continue normal operations and be able to react to contingencies that may not have formal plans. In executing these missions, combatant commands serve as the organizational unit responsible for achieving military-strategic objectives and for integrating military planning with other elements of national power to achieve grand-strategic objectives. These planning efforts produce various types of plans, but for simplicity's sake, the rest of this section only considers one notional operation plan (OPLAN). From that consideration emerges a framework for considering the assignment of forces, for determining gaps, and for reducing risk; as will be shown in a subsequent section, this framework is adaptable in the consideration of multiple families of plans. This framework may be particularly useful for a combatant command like Space Command that requires deeper integration with existing plans and will require integration with all new plans.

Within each existing theater OPLAN may exist multiple, integrated operational-level plans. For the sake of illustration, we may suppose that there is—in a common construct—a plan for each force component (land, air, maritime, and special operations). Although joint doctrine no longer advocates for a common phasing construct for such plans, phases still provide a useful method of both integrating plans and synchronizing operations. Furthermore, the plan likely designates a main effort for each phase, and each main effort is likely to determine the decisive point within their operation—the accomplishment of which will tip the scales in favor of the friendly force.

If, for example, the land force achieves the OPLAN's decisive point in Phase 3, a question we might ask is "Who achieves the land force's decisive point?" The answer may be a corps within the land force. In turn, one Division within the corps achieves the corps' decisive point. So it goes down the echelons until one might imagine that the success of the entire OPLAN hinges on one squad of infantrymen with the task to achieve their platoon's decisive point or die trying. For practical purposes, however, it is impossible to have the fidelity of an OPLAN detailed to the squad level. It is, however, entirely necessary to have a reasonable amount of fidelity in the plans of the contributing component forces.

From a space operations perspective, the central question in this scenario becomes: what space-based capabilities are necessary for the land force to achieve its decisive point? This analysis might focus on various satellite-based capabilities (communications; positioning, navigation, and timing; missile warning; intelligence, surveillance, and reconnaissance; environmental monitoring), and it might consider the requirements of the tactical elements—perhaps Army Corps or Marine Expeditionary Forces in this context—within the land force. While all of the listed space capabilities are certainly desirable for both the land force and its subordinates, analysis of the existing plan might suggest that to achieve its decisive point, the land force only *absolutely* requires satellite communications. Specifically, the land force requires satellite communications to (1) coordinate with subordinate elements and (2) coordinate with theater air forces (critical capabilities).

To say that satellite communications is a critical requirement for our friendly center of gravity, however, is not specific enough. A more specific analysis might find that several link segments, space-cyber data paths, or space-ground relay hops—conceptual data lines of communication—are mission essential. Furthermore, planners may group these data lines of communication into functional groups—for example, beam patterns of a certain portion of the UHF spectrum that cover all or part of the operating area. Such a grouping of data lines of communication is a data zone of communication.

Each of the land force's communications missions requires a data zone to be fully functional at the appointed time—perhaps one zone shared with subordinate elements, another zone shared with air forces. Within these data zones, some data lines of communication may be more essential than others—for example the data line of communication that connects the land force headquarters with the tactical main effort's headquarters. From a space support perspective, these two DLOCs may be the most critical requirements.

Figure 89 shows a notional satellite-terrestrial network during conflict. In a network, multiple data lines and zones overlap, allowing for data to traverse multiple paths toward its intended destination by traversing different, interconnected nodes. Depending on operational need, the distribution of authorities or responsibilities, and the availability of resources, data lines and zones provide useful concepts for breaking down a network into more scalable, manageable parts. By imposing a visual depiction of the satellite network over a common operating picture for ground forces, as follows, both land and space forces will be able to increase their chances of understanding how each one's operations affect the other. One may use the same process to gain a better understanding of the enemy's networks.

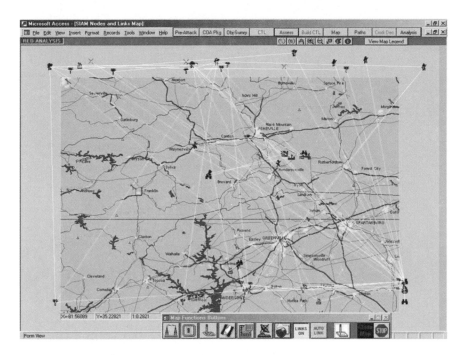

Figure 89: Notional Satellite—Terrestrial Network during Conflict. *Source:* Created
by Paul Szymanski using the Space and Information Analysis Model (SIAM)
software developed by Aegis Research Corporation under his direction.

Specialized diagrams may also aid in conceptual planning and in the
visualization of the most important data lines and zones. Figure 90 depicts
notional data lines of communication necessary to gather, process, and
disseminate imagery, missile warning, and signals intelligence across a
number of air force elements in an operational-level scenario. Because
imagery intelligence and signals intelligence involve organizations like the
National Security Agency and the National Geospatial-Intelligence Agency
(both agencies within the Department of Defense), a graphic like this may
quickly expand to account for military-strategic partners.

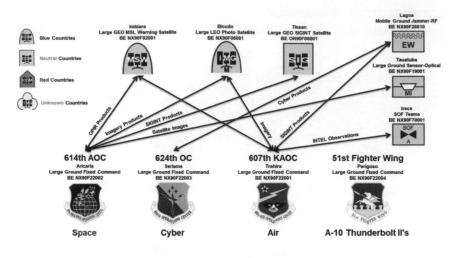

Figure 90: Notional Data Lines of Communication for Intelligence.
Source: Graphic created by Paul Szymanski.

Figure 91 functions similarly but for command-and-control communications. It distinguishes between the high data rate lines among fixed headquarters and the low data rate lines to the fielded forces. The four low data rate lines that connect the large communications satellite to the fighter wing, the jammer unit, the laser unit, and the special forces present a data zone that, because of its particular importance to mission success and its vulnerability as a low data rate zone, requires defensive measures to keep it protected—at least during the most critical phases of the operation.

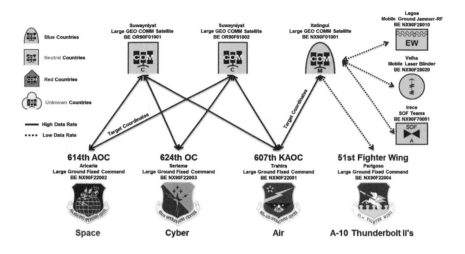

Figure 91: Notional Data Lines and Zone of Communication for Command and Control. *Source:* Graphic created by Paul Szymanski.

A depiction of data lines and zones may also expand to hint at the possibility of lines of operation. Such a depiction might be useful for conceptual operational-level planning. In the following graphic, the solid arrows on the right show that multiple special operations forces are preparing to strike multiple ground segment targets. In some sequence—perhaps simultaneously—cyber, EW, and laser units are preparing to manipulate the enemy's data lines and zones. When overlaid on a map, the physical paths of these efforts present a line or possibly multiple lines of operations. But this discussion is edging into the tactical activities of the supporting efforts. To return to the specific OPLAN in question, the main effort in Phase 3 is the land force.

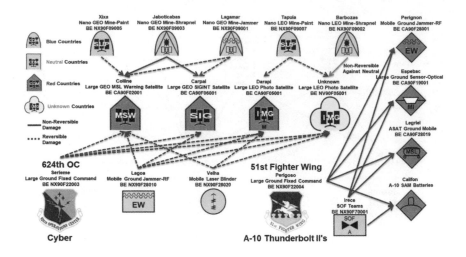

Figure 92: Conceptual Depiction of Data Lines of Communication and Lines of Operation. *Source:* Graphic created by Paul Szymanski.

For both the main effort and the supporting efforts, the establishment and maintenance of these zones is of critical importance, and from a space domain perspective, multiple options are available to aid in the accomplishment of this mission. First, established policies and robust communications planning (through theater J6 channels) will ensure proper allocation and prioritization throughout the operations. As a part of that planning, the land force will establish ground terminals to receive the satellite communications (each ground terminal providing the receptacle for a data line of communications) and ensure their protection against as many threats as possible (for example, artillery, aircraft, or sabotage). In a similar manner, Space Command would be responsible—either directly or through support agreements with another combatant command—for ensuring the security of its forces that both control the satellite in its orbit (through another data line) and manage the traffic that the satellite relays to ground stations or to other satellites (in the case of in-space data lines). Space Command forces would bear the responsibility of monitoring the data zones of communication (against either purposeful or unintentional interference) and for ensuring, to the extent possible, the safety of the satellite itself. Thus, both Space Command and the theater command—to say nothing of Cyber Command's likely involvement—share responsibility for protecting the integrity of the data zones, but each has their own particular areas of concern. Again, responsibility is likely to expand to commercial, interagency, and multinational partners.

Figure 93 depicts the flow of the logic outlined in the preceding paragraphs. While the preceding example discussed outlaying assets against decisive points based on a preexisting plan, planners might also consider the dependencies of other elements of operational design—for example, centers of gravity—as benchmarks for resource allocation. In fact, in this simplified scenario, one may argue that the land force is the operational-level center of gravity for the friendly force, that one of its critical capabilities is to communicate, and that one of its critical requirements is a particular functioning zone. That may be the case, but beginning with a consideration of decisive points in this method offers two advantages over considering centers of gravity as a starting point. First, decisive points apply to forces that may not be centers of gravity. For example, the special operations component serving as a supporting effort may not be an operational-level center of gravity, but it will still require capability allocation to achieve its decisive points, which are essential to setting the conditions for the main effort to achieve the decisive point of the OPLAN. Second, decisive points require a relationship to both time and place in a way that centers of gravity may not, which is necessary for discussing resource allocation. Yes, the Land Component will achieve its purpose at the time and place of the OPLAN's decisive point, but this is only one event in the operation, and the Land Component's status as the OPLAN's center of gravity may not guarantee that it receives prioritization outside of Phase 3. In other words, using the decisive points of the multiple operational-level forces across the entire OPLAN provides greater granularity in the effort to align scarce space resources.

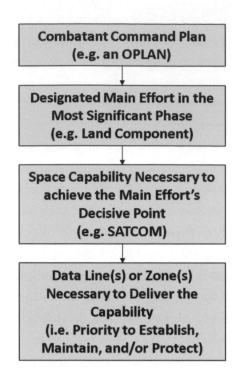

Figure 93: Single OPLAN Linkage to Data Lines and Zones for
Prioritization of Support. *Source:* Created by Jerry Drew.

In this example, the two identified satellite communications zones provide
a focal point for space operations prioritization because they are critical
to the success of the land force in achieving its decisive point. With these
requirements satisfied, the analysis should look in a similar way to the other
operational-level forces—the air, maritime, and special operations forces—and
apply resources toward the data lines and zones of communication deemed
most critical for those forces. After these requirements are met, the next step
would be to similarly prioritize line and zone allocations for other phases of
the operation. By applying resources against the operational-level planning
requirements for space operations, all combatant commands involved can
mitigate the overall risk of the associated plan and determine where external
support is necessary. Furthermore, such analysis sheds light upon capability
gaps that the institutional aspect of the military strategic can begin addressing.

Supporting Multiple Plans

Were Space Command only required to support a single OPLAN, the prioritization of support would be highly manageable, although certainly not simple. In reality, however, Space Command must contribute to many OPLANs, campaign plans, flexible deterrent options, and flexible response options—both its own and those of other combatant commands. As in the preceding single OPLAN scenario, institutional capacity presents limitations along multiple fronts (for example, available forces and equipment, training time, planning capacity, and funding) and thereby drives the need to prioritize efforts and accept risks. As with the single OPLAN scenario, the risk mitigation framework for space operations applies when considering the requirements of multiple plans.

The following graphic (Figure 94) shows notional plans for three different adversaries (Adversary A, Adversary B, and Adversary C). The combatant command responsible for Adversary A has developed three OPLANs to address possible scenarios (OPLANs A1, A2, and A3). In a similar way, the combatant command responsible for Adversary B has developed two OPLANs and a CONPLAN, and the combatant command responsible for Adversary C has developed one OPLAN and two CONPLANs. In the figure, hotter colors correspond to higher levels of assessed risk. If Adversary A is the most dangerous threat, and OPLAN A1 addresses the most likely scenario in which friendly forces will encounter Adversary A, then based on a risk assessment that considers severity and likelihood, OPLAN A1 will likely—subject to policy dictates—receive the highest prioritization. As a result, the table depicts OPLAN A1 as a red box (severe risk).

Adversary	Plan		
A	OPLAN A1	OPLAN A2	OPLAN A3
B	OPLAN B1	OPLAN B2	OPLAN B3
C	OPLAN C1	CONPLAN C2	CONPLAN C3

Risk Level Legend

■ Severe   ■ High   ■ Medium-High   ☐ Medium   ☐ Low

Figure 94: OPLAN/CONPLAN Risk Assessment for Space Operations. *Source:* Created by Jerry Drew.

To say that OPLAN A1 is the highest risk scenario, however, is not to say that all plans for Adversary A are a higher priority than all plans for either Adversary B or Adversary C. In fact, OPLAN C1 is more likely and of higher consequence (therefore greater risk) than any plan aside from OPLAN A1. OPLANs A2, B1, and B2 are of medium-high risk (light orange); OPLANS A3, B3, and C2 are medium risk (yellow); and CONPLAN C3 is low risk (green).

Within this risk construct, the military employs resources to buy down risk to acceptable levels. It may be that the available resources are used to buy down the risk of OPLAN A1 from severe to low, or it may be that available resources are used to buy down the risk of OPLAN A1 from severe to high while also buying down the risk of OPLANs C1 and A2 from high to medium. Multiple combinations are possible. In a practical sense, buying down risk equates to managing the most significant data lines and zones of communication to provide the most essential space-based capabilities within the designated plans. A complementary effort would consider the enemy's space dependencies and work to deny them those dependencies. With thorough analysis and creative thought, an approach may emerge, for example, in which the required capabilities threshold for OPLAN A2 can be largely met in preparing for OPLAN A1 and can be completely met—to within acceptable risk levels—with minor additional expenditures, with nonmateriel solutions, or through a responsibility-sharing agreement with an ally. Existing resource gaps become requirements for future programming and acquisition.

Even in the consideration of multiple plans, the prioritization framework remains very similar to that shown in Figure 93. Figure 95 incorporates the additional step of considering strategic direction, which is necessary when endeavoring to prioritize the efforts of multiple combatant commands. Second, a global military-strategic assessment prioritizes which scenarios are highest risk (OPLAN A1, then OPLAN C2, and so on). Third, an assessment of operational-level forces within each OPLAN allows for a prioritization of missions based on which components achieve the OPLANs' decisive points. Fourth, an assessment of the space capabilities required by the main efforts (and subsequently of supporting efforts) allows for prioritization of data lines and data zones of communication throughout the multiple operations. It may be that two OPLANs—OPLAN A1 and OPLAN C2, for example—could possibly occur at the same time. Combinations like this should be exercised to ensure the feasibility of simultaneously supporting both plans, the feedback from these exercises then informing the strategic approach.

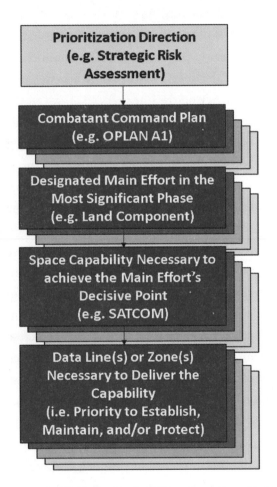

Figure 95: Prioritizing the Needs of Multiple Plans. *Source:* Created by Jerry Drew.

Insights from Data Lines and Data Zones

To be sure, the success of any combatant command plan depends on the incorporation of forces and capabilities from across the joint force and in multiple domains. From a space operations perspective, friendly and enemy forces rely on a variety of capabilities to increase the odds of mission success, all of which may be reduced in concept to their dependencies on data lines and data zones of communication. In a practical sense, then, applying limited space resources means managing data lines of communication that comprise the mission-critical functions and of denying the same opportunity to the enemy. As with lines of communication in any domain, there are three defensive ways of managing friendly data lines of communication to ensure they contribute to mission accomplishment: providing security for the line, constructing zones to increase optionality, and reducing dependency on the line.

First, a force may provide security for their lines. For historical armies, securing lines of communication has often meant posting guards, sending roving patrols, or conducting route clearance missions. This technique trades resources for security; it provides the greatest protection to the line (i.e., the lowest risk) but at a high resource cost: soldiers guarding the line cannot be used for other tasks. For space operations, there are many techniques for providing data line security. Such actions may include monitoring links for interference, employing cyber-protection teams, or employing encryption or encoding techniques. In addition, guardian satellites might be employed in the future to defend against potential satellite-to-satellite threats.

Second, a force may work to increase the number and capacity of available lines thereby forming zones. This is the most resource-intensive approach, especially if the multiple lines also require heavy security, but it also provides the greatest optionality to the force. For ground lines, this approach may mean constructing new roads, linking existing roads to form a more robust ground communications zone, complementing ground lines with sea lines as General Grant did in his Peninsular Campaign, complementing ground lines with air lines as Slim did in Burma, and/or constructing additional bases. Constructing zones of communication for the space domain may mean employing multiple ground terminals, overlapping beam patterns of the same or different satellite communication bands, employing complementary allied and commercial systems, or complementing space-based capabilities (for example, missile warning or communications missions) with data from ground, air, sea, or cyber platforms.

Third, a force may choose to reduce its dependence on its existing lines. For ground lines, this approach has historically involved reducing demand on supplied goods as General Sherman did in his campaign through Georgia or accepting risk by minimally securing a line as General Scott did in Mexico. For space operations, reducing dependencies on data lines largely equates to training to fight in space-degraded or space-denied environments. In the allocation of resources to support multiple plans, an army division or a navy destroyer squadron may find itself the lowest priority for space support. In such cases, those forces must be able to continue mission despite the lack of space-based capabilities. Determining how to best do that rests largely with the services.

Along with the three defensive ways of managing data lines (providing security for the line, constructing zones to increase optionality, and reducing dependency on the line), one must consider one offensive way: attack the enemy's data lines to induce operational friction and force his reallocation of resources as Lawrence demonstrated in Arabia. The possibilities to do this in the space domain—as hinted at in the tactics chapter—are numerous. A force may attack the ground segment, the link segment, or the space segment.

An attack on the ground segment could involve special operations forces, cruise missiles, or cyber activity. Opportunities may exist to attack an enemy's antisatellite systems or space domain awareness infrastructure before it can affect the outcome of the battle. For such a task, it is not hard to imagine the commander of Space Command taking temporary tactical control of a special forces element or of an engineer battalion especially trained to fight without space enablers.

Rather than destroying the enemy's ground segment infrastructure, however, a specially trained force—a signal company, say—could secure it and begin employing it for friendly use—the warfighting equivalent of a cavalry raid capturing a segment of enemy road and turning it over for use as a friendly supply route.

An attack on the space segment for the purposes of degrading an enemy's data lines or zones may involve, one might imagine, some of the capabilities suggested in the Defense Intelligence Agency's threat report. If the projections of that report come to fruition, kinetic attack vehicles or satellites with robotic mechanisms could be used to negate entire communications zones by either destroying or relocating satellites.* Certainly, already existing ground-based antisatellite missiles could achieve this today.

Attacking the link segment itself remains the third possibility. Negating the satellite itself or its ground control infrastructure achieves this effect, but electronic warfare assets, particularly if mobile, may provide the greatest operational flexibility.† If attacking a geosynchronous satellite, they could theoretically be placed anywhere within the satellite's zone of communications, within the zone of multiple satellites, or within a zone that is defined to include the beams of multiple satellites. In the realm of future concepts, it would be possible, with enough electronic warfare systems, to discuss the negation of entire communications zones. This challenge grows more complicated as defensive measures like beam shaping become more common, suggesting the need for future forces that are capable of getting close, inside the enemy's beam. The employment of lasers against optical sensors for blinding or permanent damage is a related one, albeit with its own unique challenges.‡ Sustained degradation of low-earth orbit imagers, for example, would require dynamic handoffs among multiple attacking elements as the satellited passed through a five-to-seven-minute window overhead, a challenging feat of both predictive analysis and of command and control.

Whether taking measures to defend one's own data lines or to attack the enemy's, there are numerous possible solutions to contribute to the

---

*   Defense Intelligence Agency, *Challenges to Security in Space* (Washington DC, 2019), 10, https://www.dia.mil/Portals/27/Documents/News/Military%20Power%20Publications/Space_Threat_V14_020119_sm.pdf.

†   Ibid., 4. The DIA reports that Russia already possesses mobile jammers for radar and SATCOM jamming.

‡   Ibid., 6. Again from the DIA, "China likely is pursuing laser weapons to disrupt, degrade, or damage satellites and their sensors and possibly already has a limited capability to employ laser systems against satellite sensors."

management of data lines. A formal analysis with the concept of the data line of communications at its center provides a useful way of thinking about all of the potential possibilities. In doing so, it becomes apparent that much of the lexicon of military theory applies even to space operations and that linking military strategic plans through operational art to tactical action is necessary in all domains. What's more, with limited resources for the conduct of global space operations, not all global planning requirements can be fulfilled by the services. This problem becomes more apparent in a consideration of space support across multiple families of plans and high-lights the value of a framework for prioritization of space assets. Deliberate management of data lines and zones, then, becomes an essential consideration for the entire joint force, for the military institutions that train and equip them, and for partners external to the US military who will contribute to future mission success.

As Space Command grapples with the myriad choices in prioritizing its limited resources, a practical approach based on an analysis of existing plans is necessary for risk mitigation in the near term. Such an approach serves as a bridging strategy to address the incorporation of space opera-tions forces and capabilities without forcing out-of-cycle rewrites of existing plans that may not sufficiently do so. Furthermore, this approach provides a way to assess capability needs as a part of the overall risk picture in a way that is complementary to institutional risk assessment processes but not dependent upon them. In considering space in the planning process—and by extension, in institutional processes—as one would consider any other domain, the joint force as a whole can become more prepared to deal with the opportunities and challenges of the space domain without being overly burdened by institutional processes.

As the linkage between military-strategic plans and tactical implementa-tion, the operational level provides the opportunity for analyzing well-de-veloped plans and considering the aspects of those plans that are most necessary for the plan's success. In the earlier example, the operational-level force depended on both tactical- and operational-level applications of space capabilities. This application centered on the concepts of data lines of communication and data zones of communication—concepts that apply to all space missions because of their dependencies on data transfer via the electromagnetic spectrum. The three historical techniques of managing lines of communication provide additional insight into the options available for making existing lines more effective, for allocating resources to future needs, and ultimately for reducing the risk to the joint force and its missions.

Conclusions

Questions of military action begin and end with strategy. As discussed in Chapter 1, these questions may be of a grand- or military-strategic nature, the latter deriving from the former. With the desired future conditions in mind, the practitioner of space warfare may begin developing the approaches to achieve those conditions. Where these practitioners sit may very well drive how they look at and attempt to solve the problems they face. Warfighting practitioners within the military strategic realm consider available means much more heavily than institutional practitioners. Theirs is a more finite time horizon with a more finite set of resources.

Importantly, the ends-ways-means framework—like the principles of war and many of the elements of operational art and design—can apply at any level of warfare. Although one may refer to a tactical plan or an operational approach, these are really strategies in the colloquial sense, smaller and simpler than their military strategic or grand strategic counterparts but nonetheless aimed at employing means to achieve objectives. The time horizon and the set of means becomes increasingly smaller and the tasks more tangible as one progresses toward the tactical level. A tactical actor knows when the unit has secured the hill; a strategic actor may not exactly know how the tactical action of securing the hill has contributed to the strategic objectives, if he is aware of the action at all.

The tactical chapter explored a number of means—orbital, electromagnetic, and surface-based—and the symbology necessary to plan tactical action both conceptually and in detail. In reality, no single tactical commander would have all of those means at his disposal. The few units available to him—based on equipment capability and the training and experience of the force—would have a set number of ways in which they could be employed. These ways are largely codified in the list of tactical tasks.

The operational level bridges the tactical and the strategic. The operational artist understands what the tactical action means to the strategic plan and continues to prosecute coordinated tactical actions to achieve operational-level objectives. These coordinated tactics may also contribute, wholly or partly, to the attainment of strategic objectives. This is the real distinction between operational art and tactics. If one takes as a starting point the tactical level (instead of the strategic level as we have done in this text), one may tend to build from the bottom up. Tactics become larger and larger, risking the false impressions that more tactics equate to operational art or that bigger tactics equate to a strategic approach. A similar danger lies in the possibility of equating targeting to tactics and believing that targeting is sufficient in itself to achieve strategic ends.

Space operations may be particularly prone to these dangers. As mentioned previously, it is tempting to believe that a rendezvous and proximity operation involving high-value friendly and enemy assets is an act of strategy. It is not. It is a tactical act that may have strategic implications, but the tactical level of war never abuts directly to the strategic level. This observation may not always be obvious, but the operational level—the level that coordinates multiple tactical actions—is necessarily always in the middle. In this case, the proximity operation simply happens to be the most visible and important tactical action (the main effort), but many other supporting activities surround it. The case is like the raid on the bin Laden compound. The SEAL team's assault on the compound itself was the main effort, but the successful accomplishment of the operation required intelligence collection, communications, cyber activity, aerial infiltration, operations in urban terrain, exfiltration, and intelligence exploitation. Coordinating all of this was the operational level that allowed for tactical action to contribute to strategic objectives.

What happens if one takes the operational level as a starting point? Here we effectively place ourselves in the middle, looking upward to the strategic level for guidance and downward to the tactical level for action. The primary danger of taking this level of war for a starting point seems to be failing to effectively line up tactical actions and strategic outcomes. Without continuous assessment, tactical actions may occur that are superfluous to the strategic goals—an inefficiency at best but a crime if lives are wasted. It is imperative in these assessments to remember that lives do depend upon space operations. To what extent remains debatable, but a GPS or SATCOM outage during combat will surely disrupt the operational tempo, delaying the attainment of objectives and incurring substantial costs.

The strategic level, then, must be our starting point and our end point when we consider how to effectively employ military power. Any military strategic planning must be firmly rooted in grand strategic objectives, and even tactical units must have a clear understanding of how their activities are contributing to higher-echelon goals. The institutional implication is that all military members must have an educational grounding in all levels of war. Codifying the collective experience of the operational force remains a goal of doctrine and a supporting effort to the institution's educational effort.

The purpose of this text has been to provide critical and creative thought about each of the levels of war vis-à-vis the space domain, how they interact, how they are confused. Explaining and clarifying has required a schema for each level, pages of military symbology, analysis of important concepts like principles of war and elements of operational art and design, and the application of theoretical ideas, historical examples, and doctrinal precepts. Each vignette drew out the interconnectedness of the levels of war, suggesting that they are not actually discrete activities but highly interrelated ones: the levels of war breathe.

Significant issues for space warfare remain, but all warfare, whether conducted by ancient Greeks or future guardians of the space force, ultimately involves a contest between adversary commanders' minds. No matter how sophisticated their military equipment, war is still about the knowledge, culture, traditions, education, intelligence, fear, and fatigue of the participants combined with the terrain, weather, political considerations, and the operational situation. Commanders employ forces and military equipment to send messages of resolve, intent, and will upon the adversary commander and his leaders. Two quotes come to mind:

"It is not the object of war to annihilate those who
have given provocation for it, but to cause
them to mend their ways."
—Polybius, *The Histories* (Second century BC)

"Why, you may take the most gallant sailor, the most intrepid
airman, or the most audacious soldier, put them at a table
together and what do you get? The sum of their fears."
—Sir Winston Churchill

Space war is real and will become even more prevalent in the near future. One cannot hide from it, and it cannot be ignored. Space is too vast and obscure to verify most treaty provisions. Antisatellite systems can readily be hidden inside innocent-looking civil or commercial satellites or hidden altogether in the vastness of space. Even seemingly innocent satellites could implement ASAT modes to disassemble another satellite. More than likely, a massive space attack will come "out of the blue" without warning. Nations that depend on space for military, diplomatic, and commercial enterprises are particularly vulnerable to these kinds of attacks. The best way to counter these potential attacks is knowledge through space domain awareness, intelligence assessments, and extensive preplanning on how to fight and win the coming war in space. There may come a time when, if a country loses a war in space, then it doesn't make sense to continue or even initiate conflicts on the ground. Since there is very little human presence in space, non-debris-causing space wars may actually preserve human life on earth.

To reiterate another theme: since outer-space warfare is so new, there is limited experience on how this kind of conflict will play out. That does not mean one should not attempt to conduct imaginative planning for this new sphere of conflict. It does mean that there will be many surprises and lessons learned after a space conflict, and the forces that appear strong on paper may actually be easily defeated by "inferior" forces with better planning or understanding of the military space environment. Furthermore, because of the global nature of space systems, military planners must also consider the political and diplomatic implications of their actions in space, if discovered.

One may easily win the space war but lose the peace afterward with new restrictive treaties and allied realignments. For this reason, it would be best to include State Department staff on space weapons development programs from the earliest start. It's conceivable that billions of dollars could be spent on a fantastic new space weapons program that the government will be hesitant to employ. Nonetheless, the most decisive weapon system employments will have to be approved by National Command authorities before use, and these authorities may have different priorities than military minds do.

The future of war and warfare has always been speculative. Many are the instances where the predictions of future wars were wrong, causing institutions to respond with equipment, training, and doctrine that may prove superfluous. Then rapid change becomes necessary. Perhaps in all wars, the fight cannot end until the institution has evolved to sufficiently meet the demands of the warfighter. In interwar periods such as the one in which the United States now finds itself, the institution will likely drive warfighting practices based on its best assessment. When conflict starts, however, the situation will reverse, and warfighters will drive institutional requirements based upon reality. A common language to discuss the levels of war, to consider the importance of space warfare to each, and to better understand this tension among competing ideas, it is hoped, will narrow the gap in understanding among all participants and allow our forces to achieve victory on behalf of the nation.

About the Authors

Paul Szymanski

World Famous Scholar and Space Expert
Is Now an Independent Consultant

The internationally acclaimed expert on space strategies and space warfare, Paul Szymanski is a major contributor to United States military space doctrine as an independent consultant. Paul Szymanski has more than 49 years of experience in all fields related to space control, to include policy, strategy, simulations, surveillance, survivability, threat assessment, long-range strategic planning, and command and control. In the past he has worked directly with the U.S. Air Force, U.S. Army, U.S. Navy and Marines as well as civilian agencies like NASA, DARPA, and FEMA. His contributions range from working for the Air Staff at the Pentagon (Secretary of the Air Force) to systems development at the Space and Missile Systems Center (SMC) in Los Angeles, technology development at the Air Force Research Lab (AFRL) in Albuquerque, New Mexico (USA) to operational field testing (China Lake Naval Test Center, California). Through this vast experience, he has a unique perspective of understanding all the divergent issues associated with each step of procurement processes.

In addition, Paul Szymanski supported the NATO Supreme Allied Commander Transformation Group and has worked in deeply covert space weapons programs (non-NRO) for 27 years, developing new concepts, analyses, and simulations while managing technical programs critical to national U.S. security. He has administered or participated in multiple architecture studies that provided senior decision-makers with long-term strategic planning road maps and were being briefed to the Secretary of the Air Force, Secretary of Defense, Joint Chiefs of Staff, Congress, and the National Security Council.

Paul is also the founder of the Space Strategies Center. His works have been featured in the Strategic Studies Quarterly of the Air University of the U.S. Air Force, Aviation Week, and he has been a frequent speaker at international conferences and exhibitions on satellite and space missions throughout the world. In total he has presented over 130 lectures and written publications in the last few years alone. He is also publishing five books this year on the topic of outer space warfare which will serve as the foundation for modern space warfare planning and strategies.

Paul also developed and manages a private space strategy related discussion group of specially hand-picked members. The membership of this group consists of over 23,000 members and include 1,700 General Officers and Admirals from the U.S. and allied nations, over 50 current and former Under/Assistant Secretaries of Defense (including one former Secretary of Defense), Congressional House and Senate staffers, members of White House and National Security Council staffs, diplomats, as well as 58 astronauts, among others.

Paul Szymanski is available for lectures, seminars, and executive consulting.

Jerry Drew

Space operations professional, military theorist, and author, Jerry Drew has been shaping the dialogue on military space operations for over a decade, most recently in his collaborative effort with world-renowned space warfare expert Paul Szymanski in their book *The Battle Beyond: How to Fight and Win the Coming War in Space*.

Throughout his military career, Jerry has served in numerous operational and institutional assignments spanning a wide variety of missions around the globe. He previously held the position of lead space planner for the U.S. Army Space and Missile Defense Command from 2018-2019. In that capacity, he served as an original member of the planning team that established U.S. Space Command and is considered a founding member of the Space Force for helping plan the new service's creation. In addition to one science fiction novel, one poem, and the foreword to one textbook, he has published a dozen articles and conference papers on topics as diverse as tactics, military history, robotics, high-altitude airships, and operational art.

Jerry is currently the chief of joint space training in the Department of Joint, Interagency, and Multinational Operations at the U.S. Army Command and General Staff College. He holds a Bachelor of Science in art, philosophy, and literature from the U.S. Military Academy and a Master of Science in astronautical engineering from the Naval Postgraduate School where his thesis work focused on applied robotic manipulation using small spacecraft. Jerry is a 2017 Art of War Scholar and a 2018 graduate of the School of Advanced Military Studies. Through those opportunities he studied the Army's role in the early Space Race and the integration of space systems into operational-level planning. He is currently enrolled as a PhD student in the Colorado School of Mines' Space Resource program where he has collaborated on a variety of projects to develop the cislunar economy, and he is partnering with Kansas State University on an upcoming modern warfare textbook. He lives in Kansas with his wife and four children.